3 12.79

# Making the Weather Work for You

## A PRACTICAL GUIDE FOR GARDENER AND FARMER

James J. Rahn

Illustrations
by
Aldren A. Watson

GARDEN WAY PUBLISHING
CHARLOTTE, VERMONT 05445

Printed in the United States

**Library of Congress Cataloging in Publication Data**

Rahn, James J 1939–
Making the weather work for you.

Includes index.
1. Weather—Popular works. 2. Gardening.
I. Title.
QC981.2.R33 630'.25'15 79-15725
ISBN 0-88266-159-0 pbk.

# Making the Weather Work for You

To Monika and Christopher...
And our future farm.

# Table of Contents


INTRODUCTION

## PART I: COLLECTING WEATHER DATA

| | | |
|---|---|---|
| CHAPTER 1: | THE HOME WEATHER STATION | 1 |
| | Basic Instruments | 4 |
| | Additional Equipment | 7 |
| | Shelters | 12 |
| | Station Location | 14 |
| | Record Keeping | 16 |
| CHAPTER 2: | PHENOLOGY—<br>USING NATURE'S OWN WEATHER INSTRUMENTS | 21 |
| | History | 22 |
| | Use In Agriculture | 24 |
| | Role of The Gardener | 27 |

## PART II: WEATHER AND CROPS

| | | |
|---|---|---|
| CHAPTER 3: | WEATHER EFFECTS ON CROP GROWTH | 33 |
| | Air Temperature | 34 |
| | Plant Temperature | 38 |
| | Soil Temperature | 39 |
| | Radiation | 42 |
| | Moisture | 49 |
| | Carbon Dioxide | 54 |
| CHAPTER 4: | WEATHER AND PLANT DISEASE | 57 |
| | Basic Plant Disease Organisms | 58 |
| | General Environmental Effects | 60 |
| | Weather Effects On Some Common Diseases | 62 |

| | | |
|---|---|---|
| CHAPTER 5: | FROST AND OTHER WEATHER HAZARDS | 67 |
| | Frost and Freezing Injury | 68 |
| | Frost Protection for Home Gardens | 72 |
| | Frost Protection (Large Scale) | 75 |
| | Forecasting Frost | 78 |
| | Drought | 80 |
| | Wind | 87 |
| | Hail | 90 |


## PART III: USING LOCAL WEATHER INFORMATION


| | | |
|---|---|---|
| CHAPTER 6: | GROWING DEGREE DAYS—BETTER THAN THE CALENDAR | 95 |
| | Plant Growth and Temperature | 97 |
| | Heat Units | 97 |
| | Use of Growing Degree Days | 101 |
| | Home Gardening and Growing Degree Days | 104 |
| | Heating Degree Days | 106 |
| CHAPTER 7: | UNDERSTANDING AND USING WEATHER FORECASTS | 109 |
| | What Does It Mean? | 110 |
| | Precipitation Probabilities | 112 |
| | Watches and Warnings | 112 |
| | Who Makes The Forecast? | 114 |
| | Getting The Current Forecast | 116 |
| CHAPTER 8: | CLIMATOLOGY AS A GUIDE | 119 |
| | Frost Probabilities | 120 |
| | Rainfall Statistics | 124 |
| | Growing Degree Day Values and Temperature Data | 126 |
| | Other Climatological Data | 127 |
| | Data Availability | 127 |
| CHAPTER 9: | THE MICROCLIMATE | 129 |
| | Temperature | 130 |
| | Precipitation | 133 |
| | Wind | 135 |
| CHAPTER 10: | WEATHER AND SITE SELECTION | 137 |
| | Garden and Orchard Planning | 138 |
| | Selecting A Home Site | 141 |
| | Landscaping | 142 |

CHAPTER 11: CLIMATIC CHANGE—FACT OR FICTION? 145
Climatic History 147
Causes of Climatic Change 150
Man-Made Climatic Changes 153
The Near Future 153
Tracking Your Own Changing Climate 156

REFERENCES 159

ILLUSTRATION CREDITS 162

APPENDIX

A. Glossary of Weather and Climate Terms 167
B. Suppliers of Weather Instruments 178
C. Sources of Weather and Climate Data 179
D. Additional Reading Material 186
E. Sources of Technical Information 190
F. Weather Data and The Metric System 191
G. Use of Gravimetric Method To Determine Soil Moisture 194
H. Analysis of Home Weather Station Data 195

INDEX 203

# Introduction

***The weather exerts a tacit, though relentless, tyranny over the labor and the thought of the agriculturist.***
H.B. Wren, 1901

One of the fastest growing hobbies in the United States today is home gardening. It may well be the true "national pastime." While farm families and many rural households have always depended upon home-grown fruits and vegetables for at least a portion of their food needs, suburbanites and even city dwellers have recently discovered the joys and rewards of the home garden. A 1978 Gallup poll determined that some form of garden was maintained in 41 per cent of the households in the United States. More than 100 million Americans lived in these homes.

This rediscovery is due to a combination of events. For one thing, the rapidly increasing cost of store-bought food has made home gardening more attractive. Also, the energy crisis has contributed to a change in the leisure habits of many Americans, resulting in more "at home" activities. Gardening is a wholesome, healthy, and generally inexpensive endeavor for all members of the family. Even pre-schoolers can help with easier garden chores, and, by the time they reach elementary school age, children may even be given their own plot.

Renewed interest in gardening has prompted an outpouring of articles, periodicals, and books, covering virtually all aspects of the subject and designed to help each grower produce a maximum amount of the highest grade fruits and vegetables. Advice is given on planting, fertilizing, mulching, composting; controlling pests and diseases; harvesting and storing, for both organic and non-organic growers.

Even in the ideal garden—with a near-perfect soil, proper fertilizer applications and pest controls, and all the other "correct" gardening practices—one element remains which ultimately determines its success or failure. That element is the weather.

An old Danish proverb noted: "The harvest depends more on the year than on the field." Every garden and farm, large or small, is subject to the vagaries of nature. Weather determines when a grower can plant in the spring and when he can harvest in the fall. During the growing season, weather affects the amount of moisture in the soil, the energy (sunlight and warmth) available for developing crops, and, finally, the yield. The effectiveness of fertilizers, mulches, pest and disease controls, and virtually all cultural practices is governed ultimately by weather conditions.

With a general understanding of local weather and climatic conditions, the use of some common weather instruments, and knowledge of a few simple techniques, home gardeners or farmers can take advantage of known crop/weather relationships to improve the efficiency and success of their operations. By adjusting habits to account for local weather conditions, growers can not only obtain improved yields, but can also reduce the use (and hence lower the costs) of materials such as fertilizers and pesticides, and can curtail the waste of irrigation water.

While it is true that on a large scale little can be done at present about our weather, the old adage "Everybody talks about the weather but nobody does anything about it" is not entirely correct. There are many ways in which the grower can either modify his local climate or adjust operations to account for local conditions. Some obvious examples include irrigation during drought, the use of heaters when frost threatens, and the timing of planting and harvesting operations to take advantage of optimum weather periods. The selection of a site for a garden or orchard can be critical in areas where late spring or early fall frosts are a frequent threat. Also, a better understanding of weather forecasting "language," knowledge of local weather conditions *at the garden site* as compared with those at the nearest official weather station, and the maintenance of local weather records can aid the grower in obtaining the best possible results from his plot under any weather conditions.

It is the intent of this book to provide you, the home gardener, homesteader or farmer, some simple, basic, and practical information about weather and climate that will help you realize the fullest possible potential from your crops—be they a couple of tomato plants in a patio box, a backyard garden, a few acres of vegetables, or hundreds of acres of grain crops or fruit trees.

## A NOTE ABOUT REFERENCES

Footnotes in this book are indicated by asterisks* and are located at the bottoms of pages. References to other books or to periodicals are indicated by numerals[1]; the references for each chapter appear at the end of the book.

# Part I: Collecting Weather Data

# 1
# The Home Weather Station

The heart of an effective program to use weather and climatic information in planning and maintaining your garden is accurate weather data—data for your own local area. Techniques described in this book for keeping weather records and for using basic weather data in guiding your day-to-day operations, such as maintaining a watch for potential frost conditions or computing growing degree days, can all be accomplished using data available through your local newspaper or your favorite television or radio station. However, such information may be based on measurements taken at a weather station miles from your farm or garden plot, and may be quite different from conditions at your site. Significant differences are likely to occur in the overnight minimum temperature and daily rainfall—two of the most important weather elements in gardening.

An obvious way to avoid this problem is to have your own weather station and collect your own data. A home weather station will provide *local* information which can be used in helpful and profitable ways. With local temperature data you can calculate your own average temperature, growing degree days, growing season length, and probable frost dates in the spring and fall, and be alert to local frost dangers. Rainfall data will determine the need for irrigation, and soil temperature will guide your spring planting operations.

Having your own weather station may seem, at first, to be impractical for the ordinary backyard gardener or homesteader—either too costly, too complicated, or both. But, the home weather station can be quite simple and inexpensive. For a modest investment of under $25 you can purchase instruments to measure the daily high and low (maximum and minimum) temperature and the daily rainfall—probably the two most important weather measurements.

## Weather Elements

If you desire, you can also make your weather station elaborate and expensive, spending hundreds or even thousands of dollars for equipment. In determining the scope of your particular setup, you should keep in mind which weather elements are of prime concern to agriculture. Of great importance, are temperature and rainfall.* Both air and soil temperatures are critical to the growth of crops and their measurement should be given high priority. If you must choose between the two, then air temperature data should be obtained first. The air temperature values can, in turn, be used to calculate growing degree days as discussed in Chapter 6, and can also be used to make a rough estimate of the soil temperature.

Moisture conditions depend, obviously, on rainfall, but also on such factors as evaporation, the initial soil moisture content, and the water-holding capacity of the soil. Rainfall is the simplest and least expensive of these parameters to measure, and of greatest value to the grower. Knowing the amount of rain that has fallen during the past day, week, month, or any other particular period, will provide an excellent guide to the general moisture conditions of your soil and the need for supplementary irrigation. This is particularly true after you have had a few years to get to know your soil moisture characteristics.

In addition to temperature and rainfall, other weather elements important to agriculture include solar radiation, humidity, dew, soil moisture, and wind. (These are all discussed in Part II of this book.) Instruments for measuring these variables are generally more expensive than a basic maximum/minimum (max/min) thermometer and raingage, but they will provide useful and interesting data.

*Household and max/min thermometers.*

*Solar radiation is probably the most important factor in plant growth (see Chapter 3), but its measurement is costly and most people are unfamiliar with its use. Even official measurements are few and far between. The use of temperature as a measure of both heat and solar energy available to the plant has proved to be a quite useful and practical substitute.

## Selecting Equipment

It is not, of course, necessary to purchase all your instruments at once. In fact to do so could prove to be almost overwhelming to some growers, suddenly faced with more weather information than they know how to handle. A better way would be to acquire some basic instruments and then, as you become familiar with the data they provide and learn to use it, other pieces of equipment can be added to your station.

In addition to selecting initial equipment, you should also decide whether you want to invest in expensive recording equipment, or begin with basic non-recording instruments. Non-recording types must be read directly and the data entered by hand into some type of record book. In general, only current and instantaneous data are obtainable. If you didn't look at the thermometer at 10 A.M., for example, you don't know, and will not be able to determine from a regular thermometer, what the 10 A.M., temperature actually was. Also, when measuring precipitation, you'll get the total amount of rainfall since your last measurement, but you won't be able to determine the amount that fell during any given time within that period. One exception to the rule that non-recording instruments provide only current data is the max/min thermometer, which will give the highest and lowest temperatures reached since the last reading and resetting, although you won't know at what time they occurred.

Non-recording equipment costs less than recording equipment. Also, non-recording instruments require virtually no maintenance and in general tend to be slightly more accurate than recording equipment, although this latter factor is usually not of critical significance.

The major feature of recording instruments is that they provide a permanent, continuous record of whatever weather element you are measuring. Generally, the data are recorded as continuous lines on charts of varying time-length. If you, or a family member or friend, cannot be at your station to read your instruments at the usual time, the data are not lost. Many recording instruments use a clock-drive or battery power to eliminate the need for a power source and are thus quite mobile and useful in remote areas. Most can be obtained with small chart drives that run unattended for periods of a day, week, or month. Other drives may be ordered with other periods. The main disadvantage of the recording instruments is their price, which may be as much as five to ten times, or more, the cost of a non-recording instrument measuring the same weather parameter.

It would probably be best to begin with some basic non-recording instruments, and then, as you see the need or desirability of adding recording equipment, to expand your weather station gradually. For gardeners and farmers who already have some basic instruments, recording equipment might be your most logical next purchase.

The most important consideration in choosing instruments should be which weather elements are of primary importance to you. A cranberry grower or an orchardist, for example, might decide their most important purchase should be a minimum thermometer, perhaps with an alarm system attached to warn of critically low temperatures. A vegetable grower in a rather arid region of the southwestern United States might purchase a raingage or a soil-moisture meter to guide his irrigation practices.

As a general rule, two instruments—a max/min thermometer and a raingage—are the most basic and useful pieces of equipment a grower can purchase for his home weather station. These should form the nucleus of any station, no matter how elaborate or expensive it might eventually become.

# BASIC INSTRUMENTS

## Non-Recording Instruments

***Thermometers.*** A common window thermometer is useful for instantaneous temperature readings, but a reliable max/min thermometer can be purchased for only a few dollars more and is of more value. Besides, there are probably few home owners who don't already have one or more standard thermometers.

There are three basic types of max/min thermometers—the "official" weather bureau type, the Six's type, and various dial types.

The weather bureau variety consists of two separate liquid-in-glass thermometers mounted horizontally on a support. The maximum thermometer has a constriction near the base of the glass tube, which allows the mercury inside the bulb to flow past as the temperature rises, but, when kept in the horizontal position, prevents the mercury from flowing back to the bulb as the temperature falls. Thus, the upper end of the column of mercury remains at its highest point until the thermometer is reset—accomplished by vigorously spinning the thermometer on its support. The minimum thermometer usually contains alcohol, and is set on a slight angle with the bulb end downward. A small metal index rests inside the upper end of the column of alcohol and is pulled downward by the retreating liquid when cooling occurs. As the temperature rises, the index remains at its lowest point until reset with a magnet. The total cost of the two-thermometer set and special support is about $40 to $55.*

*Prices given in this book are offered for general guidance only. Variations occur among manufacturers and sellers, and general price changes are likely with time. The latest prices can be obtained by requesting catalogs from companies listed in Appendix B.

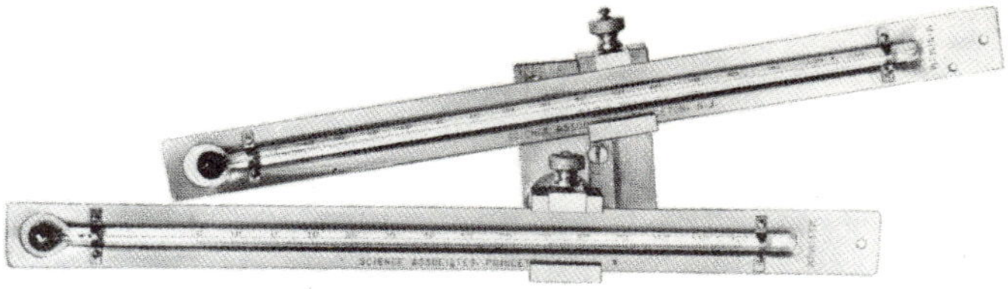

*"Weather bureau variety" max/min thermometer.*

A less expensive max/min thermometer is the Six's variety, consisting of a U-shaped glass tube partially filled with mercury, with a metal index outside the end of each mercury column. Rising temperatures will cause the mercury to move up one side of the tube, pushing the index to the highest value reached during any given period. As the temperature falls, the mercury column retreats, moving up the other side of the tube, where a rising column indicates lower temperatures, and the minimum index is pushed to the lowest temperature for

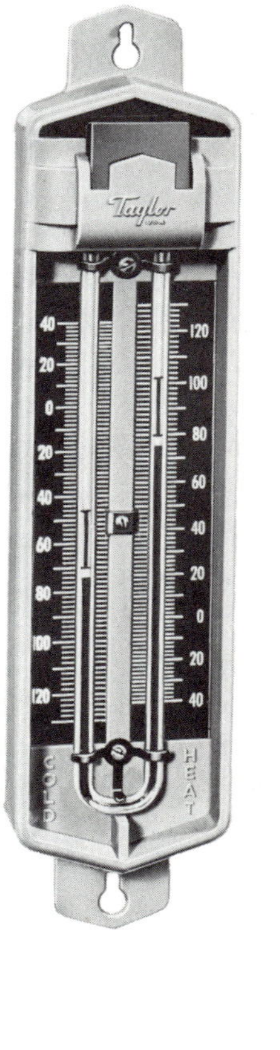


*Six's variety, u-shaped max/min thermometer.*

the period. The indices are reset with a magnet. The cost of this max/min thermometer is about $15. No special support is required. The Six's thermometer is quite accurate and very popular.

The third type of max/min thermometer is the dial type, in which three pointers are used to indicate the current temperature and the maximum and minimum temperatures since the previous setting. These instruments appear similar to dial-type thermometers used in roasting meat. The pointer indicating current temperature moves the other two pointers as it varies from its high to its low readings. These latter two pointers are then read and reset by hand. Dial-type thermometers are not as widely used as the two previously mentioned varieties and generally cost from about $35 to $60. (A lower-priced version, costing under $10, has been advertised in some garden and plant publications. If it proves to be accurate and durable, it would be an alternative to the Six's type for home gardeners.) No special mounting is required for these instruments. For all three types of thermometers, however, some form of shelter is necessary, as discussed in the section on shelters later in this chapter.

*Raingages.* A variety of non-recording raingages is also available, ranging from the standard weather bureau 8-inch gage, used at most climatological stations, to an assortment of smaller, usually plastic or plexiglass, models. The standard 8-inch gage consists of an 8-inch diameter collector and funnel, which fits over a large overflow can. Inside the larger can is a smaller brass collection tube which catches the rain from the funnel. A wooden or plastic measuring stick allows measurement of precipitation to the nearest one-hundreth of an inch (0.01"). Total capacity of the gage is 20 inches. The cost of the standard 8-inch raingage, with support, is about $90.

Many smaller and considerably cheaper raingages are on the market. There are plastic

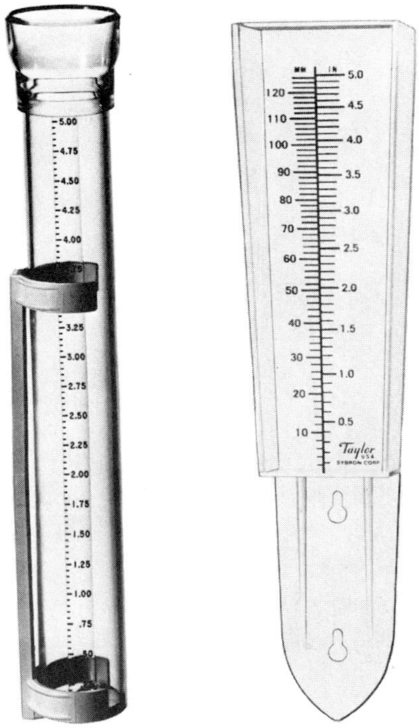


*Typical, inexpensive raingages, suitable for home gardening or farming.*

(or plexi-glass) varieties which are smaller versions of the 8-inch standard gage, the wedge or "farm" raingages, and others of various sizes and shapes. While the accuracy of some may not be quite as good as that of the standard 8-inch variety, they are sufficient for the home gardener or farmer. Most of these gages can be attached directly to a fence post or pole, while a few come with their own support. One note of caution—some of the plastic models may crack if water sits inside and freezes. If your only interest is in recording rainfall during the gardening season, this may not be an important consideration. For more serious amateur weathermen, it would pay to get only models which can be used year-round. If there is any question about their suitability for winter use, check with the manufacturer.

A remote-reading raingage, which provides a digital readout of rainfall at some distance from the gage, is available for about $40 to $50.

## Recording Instruments

***Thermographs.*** Recording versions of the basic meteorological instruments described above are readily available. Thermographs, which give a continuous reading of temperature, can be obtained for from $65 up to several hundred dollars. These instruments generally consist of a temperature sensor connected to a pen arm, a clock or battery driven chart drum, and the appropriate charts. Usually, the drum rotates once each day or week. The resulting continuous temperature record allows you to determine the temperature at any time or to calculate the average temperature for any particular period. There may be a slight lag in the reaction of the pen arm to changes in temperature, depending on the chart scale and speed. While this lag or "damping" effect is usually not critical, it may affect the peak maximum or minimum values—reading slightly low at the maximum and slightly high at the minimum. For this reason, a regular max/min thermometer is recommended, even if a thermograph is being used.

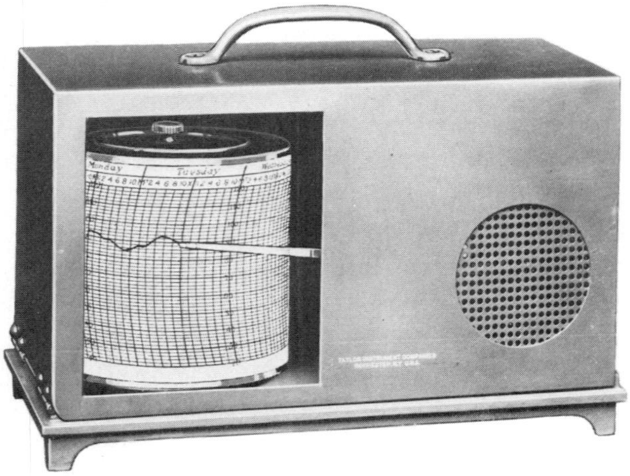

*Thermograph with pen arm and chart drum for continuous temperature recording.*

The high and low values from the max/min thermometer should be used as the true daily extremes and can be used to keep a check on the calibration of the thermograph. The thermograph record is most valuable for determining the times of the highest and lowest temperature as well as the general shape of the diurnal (daily) temperature curve. Hydrothermographs, providing recorded values of both temperature and humidity, are also available at somewhat higher prices.

***Raingages.*** Recording raingages generally cost at least $300. There are two basic types—the weighing bucket, and the tipping bucket gages. In the weighing bucket models, rainfall or melted snow is funneled into a large bucket which rests on a scale-type support. An increase of water in the gage (increased weight) drives a pen arm which records the reading on a moving drum chart.

With the tipping bucket variety, two small "cups" alternately fill with each 0.01 inch of precipitation that falls. When one cup is full it dumps, moving the second cup into position and closing an electrical contact, which moves a pen arm causing a mark on a moving drum chart. Each 0.01 inch of precipitation is thus recorded as a "blip" on the chart.

Both types of raingages provide a continuous record of rain or snow, allowing calculation of rainfall *intensity* (the rate of rainfall over a particular period of time), as well as rainfall amount. Knowing the rainfall intensity gives a good indication of the amount of rain that soaked into the ground. If an inch of rain falls over a period of 12 hours, for example, much more of the rain will soak into the ground than when an inch falls in an hour. In the latter case, much of the water will run off the surface and may even cause some soil erosion. A good fertile soil with high organic matter content will absorb more of a heavy shower than will a hard clay soil. Also, a good mulch will help prevent excessive runoff and allow more rain to seep into the soil.

*Recording rain and snow gage.*

# ADDITIONAL EQUIPMENT

There is quite an array of additional weather instruments. How elaborate and expensive a home weather station you set up will depend on your own interest, special needs, and finances. Equipment discussed in the following paragraphs will provide useful information to growers.

*Mercury-in-glass soil thermometer.*

## Soil Temperature

While there is no set order in which you should purchase your equipment, one of the first items you may want to consider after obtaining the basic max/min thermometer and raingage, is a soil thermometer. Soil temperatures are critical in the spring for seed germination and continue to be important throughout the growing season. They affect the existence and spread of soil borne diseases and pests, and control the chemical and biological activity within the soil. Soil temperatures may be obtained from either of two basic thermometer types—regular mercury-in-glass models, or dial types. The latter cost between $10 and $20 while mercury-in-glass models are more ex-

*Gardener's soil thermometer.*

pensive, costing from about $30 to $65. Of even more value, but rather expensive, are soil max/min thermometers. These are generally of the dial type, resembling the dial-type max/min thermometers used for air temperature measurements. The soil temperature version sells for about $80. Recording soil thermographs are also available, ranging in price from about $225 to more than $500. The more expensive models allow you to record temperatures from two or three locations—for example, soil temperature at three different depths or perhaps soil temperature at two depths plus the air temperature within the garden.

## Soil Moisture

Soil moisture is also of great interest to farmers and home gardeners. Its measurement, however, is not as simple as is that of the soil temperature. The most popular soil moisture meters are those which take account of the effect of moisture on electrical resistance. Various types of resistance-block meters are available. These blocks are placed at desired levels within the soil and the measured electrical resistance is then related to actual soil moisture values. The equipment costs about $100 or more.

Low-cost moisture meters have become available, designed primarily for use with house plants. These compact, lightweight units consist of a probe which is inserted into the soil and a dial on which the moisture content is registered on a simple scale of 1 to 8, or 1 to 10. The higher the number, the wetter the soil. Usually an accompanying booklet lists the preferred moisture level for many popular house plants. While designed mainly for inside use, there is no reason why this instrument cannot be used outside in flower or vegetable gardens. In fact, I've done just that—using one of these meters to test the moisture content in my garden soil, at a time when irrigation was being considered because of rather dry weather. The moisture meter proved helpful, indicating that while the top inch or so was quite dry, at several inches below the surface there was considerable moisture remaining in the soil.

That simple test, repeated at various locations in the garden, prevented an unnecessary watering. Each user will have to determine the scale reading at which irrigation is necessary for his particular soil, but this instrument does appear to be a useful gardening aid. Most models cost between $10 and $15. A combination water and light meter is also available at a slightly higher price.

The most accurate method of determining soil moisture actually requires little equipment—mainly a trowel, a container and an

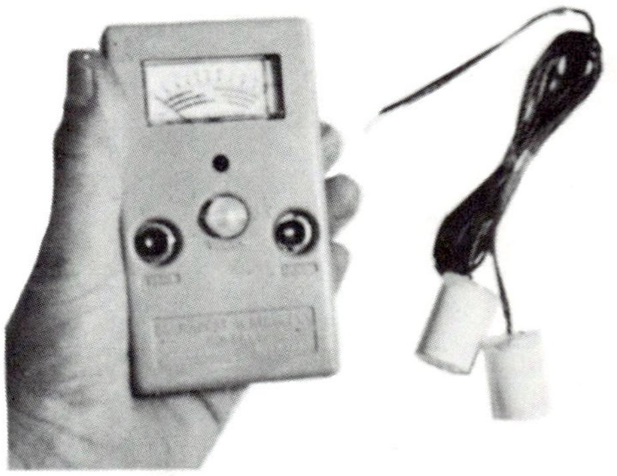

*Low-cost moisture meter with probes.*

accurate and sensitive scale. A sample of soil is taken from the garden and weighed. The soil is then oven-dried for a few hours at about 212° to 230° F. and weighed again. The difference in weight between the first and second weighings represents the weight of the moisture in the soil. This value, when divided by the weight of the oven-dry sample, gives soil moisture as a percentage, by weight, of the dry soil (see example in Appendix G). This technique is known as the *gravimetric* method of soil moisture determination. Its main drawback is that it is rather messy and time consuming.

## Evaporation

The measurement of evaporation, in conjunction with rainfall measurements, will provide an accurate estimate of net water gain or loss in your soils. With a bit of experience you will be able to make a better estimate of the need for irrigation than from rainfall records alone. Simple evaporimeters, consisting of a glass cylinder with a disc of porous paper on the open end, cost about $15. This instrument provides a useful indication of the evaporative "power" of the atmosphere, which can be related to water evaporated and transpired from a growing crop. Recording evaporation gages, in which changes in the weight of water contained in a shallow pan are converted to evaporation losses, sell for about $400. Official evaporation pans and accessory equipment used at climatological stations cost $600 or more.

## Relative Humidity and Dew

Relative humidity is an important factor in estimating the onset and severity of certain plant diseases. Quite accurate readings of relative humidity can be obtained through the use of hand-held sling psychrometers, which sell for about $10 to $20. Basically, these instruments consist of two regular thermometers set in a special holder with a handle for whirling the thermometers through the air. The bulb of one of the thermometers is covered with a muslin wick, which is wetted before use, and the instruments are then whirled through the air until both thermometers reach their equilibrium temperature—that is, until continued whirling produces no further changes in the temperature. The difference between the dry-bulb and the wet-bulb temperature is converted into relative humidity through the use of appropriate conversion tables provided with the instruments. The dewpoint temperature can also be obtained using the wet-bulb and dry-bulb readings. The dewpoint temperature is the temperature that the air must be cooled to before dew forms. It is a useful parameter in frost forecasting.

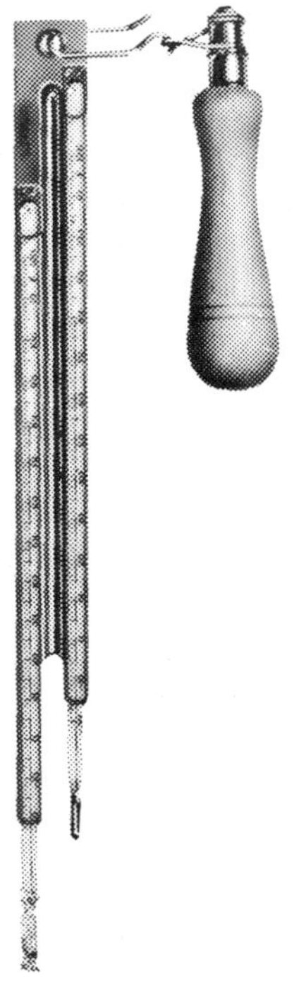

*Hand-held sling psychrometer.*

Aspirated psychrometers, in which a tiny electric fan is used to blow air past the thermometer bulbs (avoiding the necessity of whirling the instrument through the air), cost from $75 to $300 or more. The expensive models use electrical semi-conductors (thermistors*) to measure the temperature, rather than liquid-in-glass thermometers. Recording hydrothermographs which continuously record relative humidity and temperature cost at least $150. The humidity readings with the hydrothermograph may be somewhat less accurate than those obtained with a sling psychrometer, but offer the advantage of a continuous record. The sling psychrometer can be used to check the calibration of the recording instrument. Remote reading relative humidity dials, using an electrical cable to connect the sensing element out in the field to a dial inside the house or other shelter, cost about $400.

Gages are also available to record and measure dew, although their usefulness to the average gardener or farmer seems limited. Non-recording dew gages cost about $50, while recording models sell for about $300 or more.

*Wind speed and direction indicator.*

## Wind

Wind measurements are of general interest to many people and are important to certain agricultural operations such as spraying and dusting. In addition, changes in wind speed and direction can give important clues to impending changes in the weather. Simple weather vanes at various prices can be used for indicating the wind direction by sight, while hand-held wind speed meters are available for about $25. For more accurate measurements, however, wind equipment in which the speed and direction are indicated on appropriate remote-reading dials should be considered. Various types of wind sets are available, ranging in price from about $85 to more than $300. Recording wind instruments are available at higher prices.

## Solar Radiation

Equipment to measure solar radiation is available in a variety of models. Most economical units cost about $90 and more expensive

*Thermistors are resistors that are particularly sensitive to temperature changes. They may be in the shape of small beads, rods, discs, or plates. Depending on their use they may be a cm or two in length (or diameter) down to a few mm. When used as a temperature probe, they may be sheathed in stainless steel for protection.

units range in price to about $1000. Although not widely recorded, solar radiation is the ultimate source of energy for plant growth and development and much study needs to be done on solar/plant-growth relationships.

A simple barometer is this one with a 5¼-inch dial and heavy brass case.

## Air Pressure

Barometers, while not instruments with direct application to agriculture, are nevertheless quite useful to the amateur weather "buff" in helping to anticipate future weather conditions. A variety of styles and prices are available. Some models sell for only a few dollars, but for the serious amateur it pays to purchase one of the more accurate and sturdy types, versions which cost from about $20 up to several hundred dollars. Good models are available in the $30 to $70 range. Recording barographs cost approximately $75 to more than $500.

For those interested in acquiring an "instant" weather station, complete station setups are available, costing about $80 for a simple, basic, and practical version, up to several thousand dollars for professional models.

*Another accurate, sturdy barometer is this Navy type with a 5-inch dial.*

*This recording barometer keeps track of barometric pressure changes continuously on a seven-day chart.*

Because of the variety and frequent changes in both styles and prices, those interested in obtaining weather equipment should consult one or more of the companies listed in Appendix B. Obtain a catalog from at least two of the suppliers and compare prices. Some of the more common instruments, such as max/min thermometers, raingages, soil thermometers and moisture meters, are often available at large department and discount stores, garden centers, and through seed and plant-supply catalogs.

## SHELTERS

The final item of your home weather station should be some type of weather shelter. Shelters protect instruments from radiation (sunlight), wind, moisture, and any other source of physical damage; and they prevent erroneous readings caused by these weather elements.

The primary use of shelters is to house thermometers. In order to obtain true air temperatures it is necessary that the thermometer be protected from sunlight and from moisture. Radiant energy from the sun will obviously produce a reading too high (try laying a thermometer in the sun), while a wet thermometer will read below the true air temperature because of evaporational cooling.

The standard climatological shelter (sometimes referred to as the cotton region shelter) can be purchased for about $130. Made of wood and painted white (to reflect sunlight) it has a double roof to reduce radiation effects, and louvered sides and flooring to permit good air circulation. For maximum benefit it should be kept clean and not allowed to fade in color. The legs of the shelter may be either wood or metal. The center of the shelter is usually placed about 5½ feet above the ground. This type of shelter can be used for any of the various max/min thermometers, as well as for thermographs or hydrothermographs.

A smaller shelter, designed mainly to hold the Six's type max/min thermometer, is also available. This version is generally mounted on a wall or post, preferably facing north, and should be shaded from direct sunlight. It sells for about $40. Various other types of shelters are available for specific purposes, but, for the average home weather station, the two types described above are generally sufficient. Descriptions of other shelters can be found in catalogs supplied by the companies listed in Appendix B.

*A standard weather instrument shelter.*

For do-it-yourselfers, it is quite easy to build your own shelter, keeping in mind the following basic requirements:

1) Instruments must be protected from direct or reflected sunlight.
2) Instruments must be kept dry.
3) An adequate airflow must be permitted to prevent heat build-up.
4) The thermometer should hang "free"—that is, it should have no direct contact with the sides of the shelter in order to reduce heat conduction from the shelter to the instrument. This is best accomplished by placing the thermometer on a wooden crossbar within the shelter.

*Weather Instrument Shelter*

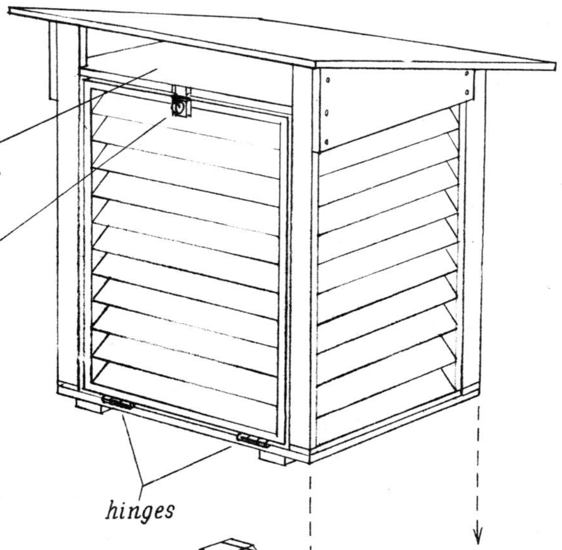


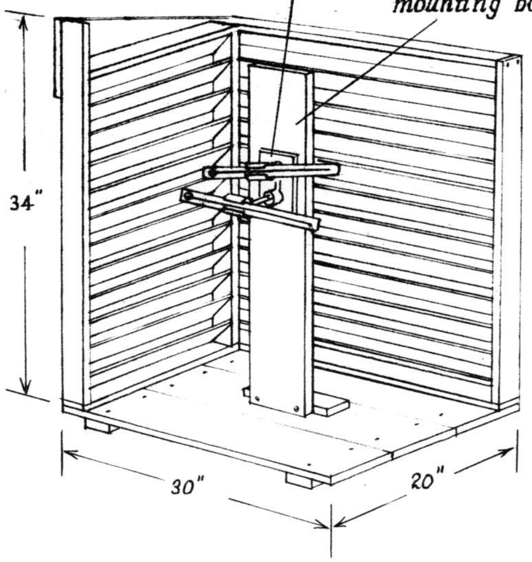


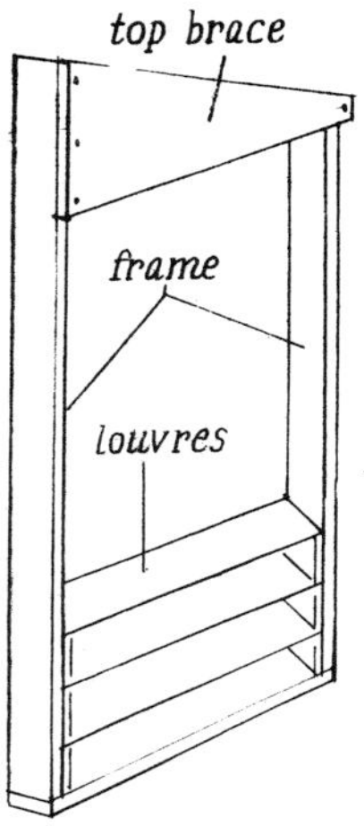


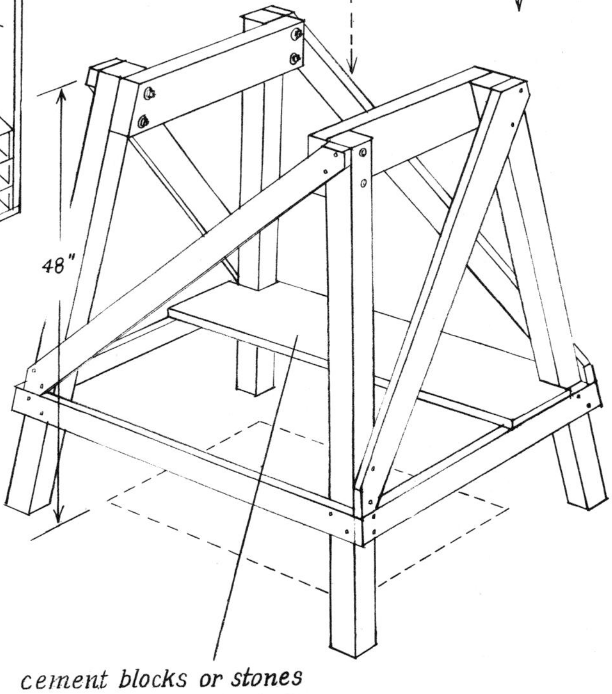


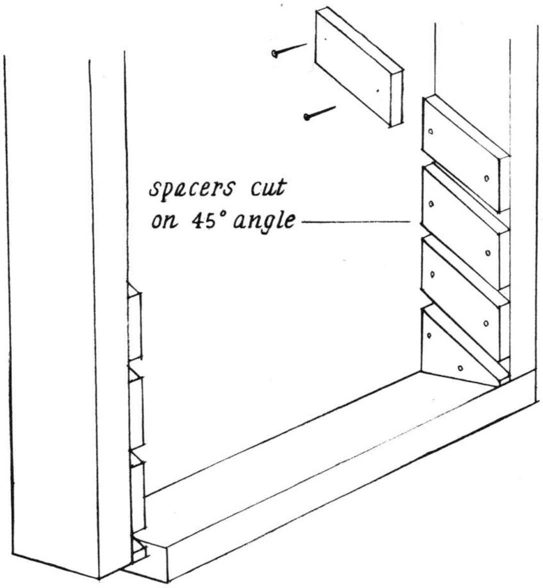

Wood is generally the best material to use for the shelter because of its low heat conductivity. A white paint will provide good reflectivity of sunlight, while louvered or slatted sides and flooring will permit a free flow of air.

While from a very strict and technical viewpoint there are certain criteria which must be met in a shelter, a simple homemade housing will generally provide temperature readings that are quite accurate for all practical purposes for the layman meteorologist. Consistency is as important as the type of shelter. That is, once you have a shelter, keep using it, unless it is quite obviously inadequate. Frequent changes in either the type or location of the shelter will cause more problems than temperatures that may be a degree or two too high on a sunny afternoon because of some radiation error.

Most other instruments do not require a shelter. A raingage, obviously, would be useless if covered. Soil thermometers, depending on the type, are usually fully exposed, although with the dial type a small can may be set over the dial to prevent moisture from working into the dial face and causing a fogging problem. Also, a covering will keep ice and snow off for easier reading during the winter, if left in the garden. The can should be large enough to just fit over the dial, to minimize shading of the soil.

Recording instruments, such as a thermograph, require a shelter similar to that described above. The recorder part of remote-reading instruments can be kept in your house, garage, or shed, with a cable running outside to the sensing element. This would be the case, for example, with wind instruments, where the vane and anemometer are placed on the roof and the wind direction and speed dials are located at some convenient place indoors. Generally, when purchasing an instrument, you will get specific directions concerning its proper exposure and location.

## EQUIPMENT LOCATION

Once the equipment has been selected and purchased, the remaining decision is where to locate it. Each backyard or homestead presents its own unique situation when considering the convenience and ease of accessibility of the equipment. However, there are some general guidelines to follow.

Raingages should be located away from tall objects such as trees or walls of buildings and should be level. If placed on sloping ground, a piece of wood or other material should be used to level the gage opening in a true horizontal position. If possible, avoid areas where the wind "funnels through" (such as between buildings), since a strong wind flowing past the gage reduces its accuracy. Place the raingage close to your garden, if you are mainly interested in rainfall at that location.

Weather shelters should be placed in an open location where wind flow is not restricted. Usually, they are placed over a grass surface. If you are looking for a record of conditions *in* your garden or cropland, however, then place the shelter in, or next to, them. If your land is hilly, you may find it interesting to compare temperatures in the low spots with those higher up. In any case, avoid concrete or asphalt surfaces—unless you are specifically interested in the temperature over asphalt! The door or open side of the shelter should face toward the north to help reduce radiation errors.

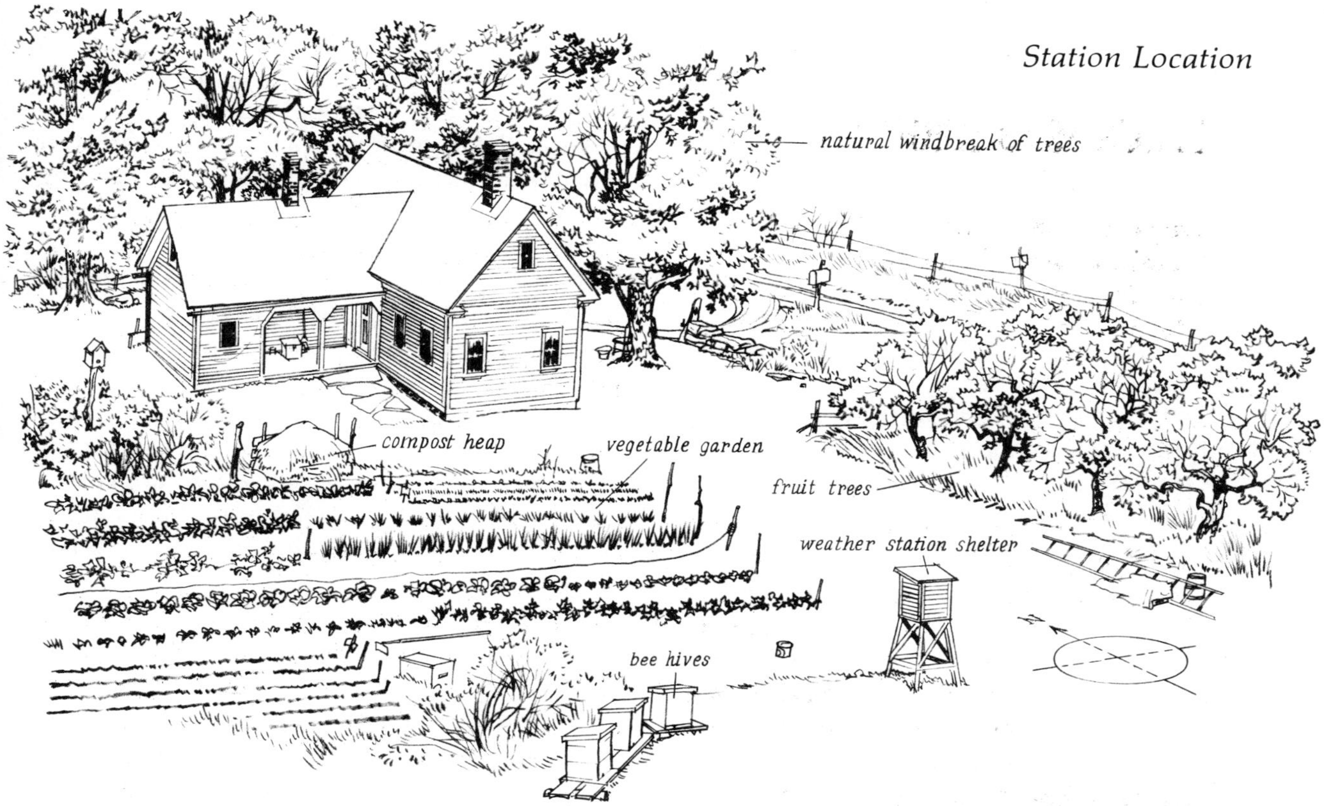


Evaporation equipment should be located close to your plot and exposed to the same general conditions as your crops. Wind instruments are generally placed on a pole or rooftop above surrounding obstacles, if at all possible. In some cases, however, you may be more interested in wind flow down at crop level, such as for guiding spraying or dusting operations.

Again, consistency is important. You will be using your weather data to develop certain environment/crop-growth relationships for your own particular site. A temperature reading that is off by a degree or two, or a rainfall value that might be off a few hundredths of an inch will generally not be as important as differences caused by changing the location of your instruments. Unless it is obvious that a change is necessary (tree branches overhanging a raingage, for example) pick a good spot for your weather station and keep it there.

# RECORD KEEPING

You have your instruments, you've bought or made a thermometer shelter, and you've selected a site for your weather station. Now you are ready to begin your own climatological record. What should you keep records of—and how?

Probably the most important rule-of-thumb is: Don't trust your readings to memory—write them down at once. You may have a good memory, but keeping a maximum temperature, a minimum temperature, the current temperature, total rainfall, and soil temperature in your head for very long will undoubtedly lead to errors. So when you read your instruments—write the values down! Carrying a small note pad and pencil with you is the best hedge against a memory lapse.

In general, taking readings of temperature and rainfall once-a-day is quite sufficient. Record the maximum and minimum temperature since your last reading and also the current (observation) value. Don't forget to reset the indices on your thermometer! Measure the rainfall since your last observation and then make sure the gage is completely emptied. Evaporation readings and maximum/minimum soil temperatures are also usually recorded on a daily basis. If you just have a regular dial type soil thermometer you may want to read it several times each day to see how it varies with time.

While taking your readings each day at midnight would give you true daily values for each calendar day, such a practice is not necessary and is followed only at weather stations open 24 hours a day. It *is* important that you read your instruments at the same time each day, if at all possible. The majority of climatological stations take readings about 7 A.M. or 7 P.M. There isn't anything special about those times except they seem to be convenient for most observers. When you take your readings will depend on your personal schedule—but be consistent! Of course, you can look at your instruments as often as you desire, but keeping records on a daily (that is, 24 hour) basis is the standard convention and will provide the most useful data. Keep in mind that if you take your temperature readings in the evening the maximum and minimum temperature will most likely (although not always) have both occurred that calendar day. If you read your thermometer in the morning, however, in most instances the high temperature for the 24-hour period will have occurred the previous afternoon. Always document the time when readings are taken.

Recording instruments give a continuous record of a particular weather element and can be very useful. Thermographs, for example, can show when the highest temperature occurred, or tell you how long your garden was subject to freezing temperatures overnight. With a recording raingage you can determine the rainfall intensity—or, how "hard" it rained. Even so, your basic records should still cover a 24-hour period, since nearly all the techniques described in this book (and generally used in agricultural weather studies) for using weather data are based on daily records.

Each of you will develop your own technique for keeping your weather records. A general record sheet format is suggested in Table 1. Usually one sheet of records is used for each month. Monthly average temperatures, temperature extremes, and total monthly precipitation, snowfall and evaporation, if available, are usually summarized.

## Table 1: Monthly Weather Record

Station Name ______________________ Month ______________________ Year ______
County ______________________ Observation Time ______________________
State ______________________ Observer's Name ______________________

| | Air Temperature | | | | Soil Temperature | | | Precipitation | | | |
|---|---|---|---|---|---|---|---|---|---|---|---|
| Date | Max | Min | Mean | Obs. | Max | Min | Obs. | 24-Hour Total | New Snow | Snow Depth | Remarks |
| 1 | | | | | | | | | | | |
| 2 | | | | | | | | | | | |
| 3 | | | | | | | | | | | |
| 4 | | | | | | | | | | | |
| 5 | | | | | | | | | | | |
| , | | | | | | | | | | | |
| , | | | | | | | | | | | |
| , | | | | | | | | | | | |
| , | | | | | | | | | | | |
| , | | | | | | | | | | | |
| 31 | | | | | | | | | | | |
| Monthly Mean: | | | | | | | | | | | |
| Monthly Extreme: | | | | | | | | | | | |

**NOTE:** Obs. is the temperature at the time of observation.

On your record sheet you should have room for any remarks you may wish to make (such as the occurrence of thunderstorms, hail, or frost), as well as for any additional data, such as wind and evaporation, which you may record. If you have soil thermometers at more than one depth, keep the record of each depth separate. The mean temperature is simply the average of the daily max and min, or max + min ÷ 2. In measuring total precipitation (usually to the nearest 0.01 inch) both rain and/or melted snow are included. If there is snow in your gage, take the gage inside and melt the snow (or ice) to determine the actual moisture content. The moisture content of snow is sometimes referred to as the *water equivalent*. You will find that one inch of snow usually contains roughly 0.1 inch of water, but this snow/rain ratio will vary depending on the temperature. Generally, the colder it is, the drier the snow.

Snowfall (new snow since the last observation) is usually measured to the nearest 0.1 inch at official climatological stations, but for practical purposes may easily be measured to the nearest .25 or .50 inch. Total snow on the ground is measured to the nearest whole inch. Note that *snowfall* and *snow depth* are not always equal. For example, if you had 5 inches of snow during the morning, but some melted during the day, and when you take your regular reading in the evening only 3 inches remain, then your new snow would be 5 inches, while the depth of snow on ground at observation time would be 3 inches.

## Table 2: Summary of Growing Season Weather

| Year | Mean April-Sept. Temperature | Total April-Sept. Rainfall | Last 32° in Spring | First 32° in Fall | Growing Season Length | Accumulated Growing Degree Days Base 40° | Accumulated Growing Degree Days Base 50° |
|---|---|---|---|---|---|---|---|
| 1974 | | | | | | | |
| 1975 | | | | | | | |
| 1976 | | | | | | | |
| 1977 | | | | | | | |
| 1978 | | | | | | | |
| 1979 | | | | | | | |
| 1980 | | | | | | | |

A word of caution: Some of the plastic rain gages may crack or split if rain water or melted snow freezes in the gage. Such gages are really only useful during non-winter months. This may not be too important for those who are only interested in growing season rainfall, but it is important to those who want a year-round record. Before purchasing one of the plastic or plexi-glass rain gages, *make sure* it can be used during freezing weather.

In addition to the direct measurements you make from your instruments, you will also want to keep certain other calculated data. These may be kept on a separate record sheet or as part of your regular monthly record. Calculated data would include such values as growing degree days (Chapter 6) and growing season rainfall, accumulated from a specific date each spring. You may also find it interesting to compute heating degree days (Chapter 6) as a check on heating fuel requirements for your home or greenhouse.

Other special types of information that may prove useful and that will help you to better define your own local climate include:

1) The last date in spring and the first date in fall on which the temperature falls to 32° F.* or lower. The period between these dates represents your freeze-free season.

2) The date on which the soil temperature reaches certain values, such as 40°, 45°, and 50° in the spring. This information will aid in planning your seeding dates.

3) The number of days each year with minimum temperature below 32°.

4) The number of days each year with the maximum temperature either below 32° or above 90°.

5) The number of days during the growing season with 0.10 inch or more of rainfall.

6) The number of days with thunderstorms and/or hail.

7) The length and dates of the longest rain-free period each growing season.

8) The number of days each winter with at least one inch of snow cover.

The above items are only suggestions, and whatever information you find useful or interesting can be calculated and recorded.

As you accumulate several years of records you can summarize your data by months, years, or growing seasons. An example of a growing-season summary is given in Table 2.

*Unless otherwise noted, all temperatures in this book are in degrees Fahrenheit (°F.). For conversion to degrees Celsius (°C.), see Appendix F.

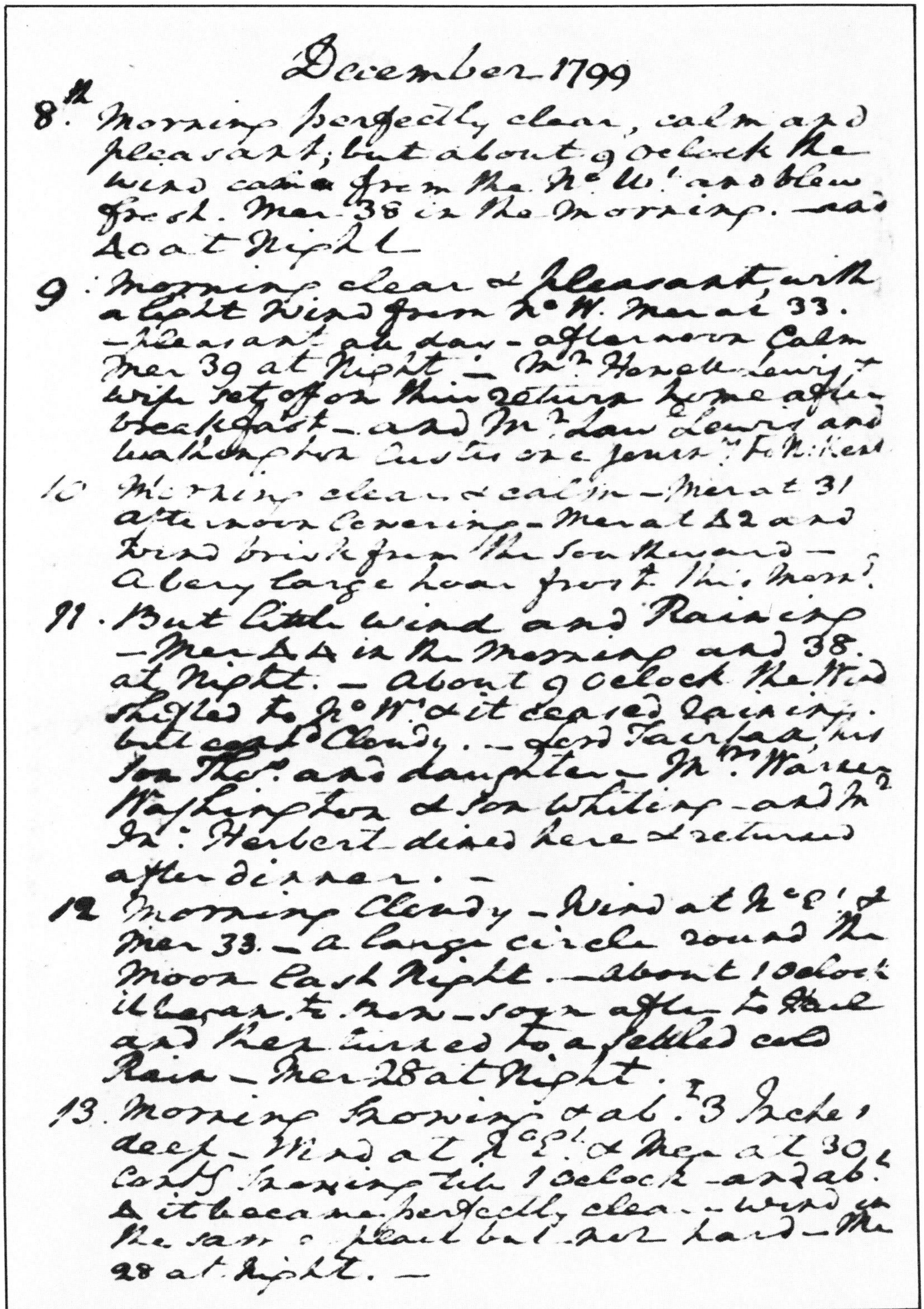

December 1799

8th. Morning perfectly clear, calm and pleasant; but about 9 o'clock the wind came from the No. Wt. and blew fresh. Mer. 38 in the Morning—and 40 at Night

9. Morning clear & pleasant, with a light Wind from N°. W. Mer. at 33.—Pleasant all day—afternoon Calm Mer. 39 at Night—Mr. Howell Lewis & wife set off on their return home after breakfast—and Mr. Law Lewis and Washington Custis on a journey to New Kent

10. Morning clear & calm—Mer. at 31 afternoon Lowering—Mer. at 42 and Wind brisk from the Southward—A very large hoar frost this Morng.

11. But little wind and Raining—Mer. 44 in the Morning and 38. at Night.—About 9 o'clock the Wind shifted to N°. W. & it ceased raining but contd. Cloudy.—Lord Fairfax, his Son Thos. and daughter—Mrs. Warner Washington & Son Whiting—and Mr. Jn°. Herbert dined here & returned after dinner.—

12. Morning Cloudy—Wind at N°. E. & Mer. 33.—A large circle round the Moon last Night.—About 1 o'clock it began to snow—soon after to Hail and then turned to a settled cold Rain—Mer. 28 at Night.

13. Morning Snowing & abt. 3 Inches deep—Wind at N°. Et. & Mer. at 30. Contg. Snowing till 1 o'clock—and abt. 4 it became perfectly clear—Wind in the same place but not hard—Mer. 28 at Night.—

*George Washington's weather diary. The final entry was made the day before he died.*

# 2
# Phenology—Using Nature's Own Weather Instruments

We have looked at a variety of instruments, each of which measures a particular weather element such as temperature, radiation, moisture, and wind. Each of these weather parameters is important and of interest to gardeners and farmers. What would really be useful, though, is an instrument that would integrate, or combine, all of these factors into an index or other measure of the overall weather conditions, particularly as they affect plant growth and development. So far, despite all his ingenuity, man has been unable to devise such a piece of equipment.

But then, does he really need to? Nature's plants and animals provide exactly that information. We need only to take the time to observe and learn from them. By noting the time of certain periodic events in the natural cycle of plant development, for example, we can obtain information about recent weather and climatic conditions which no single man-made instrument can measure.

# HISTORY

The study of these natural periodic events in the life-cycle of plants, such as the first leafing or flowering in the spring, is called *phenology,* derived from the Greek work "phaino" which means to show, or to appear. The term phenology is believed to have been first used by the Belgian botanist Charles Morren about 1853, but the observation of phenological events dates back many centuries to the ancient Chinese, who had developed a phenological calendar several centuries before Christ.[1] With the development of the thermometer in the 18th Century it became possible to relate these phenological events to temperature and renewed interest in phenological observations resulted.

In the United States, phenological records began to appear during the latter part of the 1800s. One of the earliest works was that of C. Hart Merriam in the 1880s.[2] In 1918, A. D. Hopkins developed his "bioclimatic law"[3], which states that the advance of spring is delayed by four days for each degree of latitude north, for every 400 feet of elevation upward, and for every five degrees of longitude eastward from the Rocky Mountains.

These laws are rather general and do not

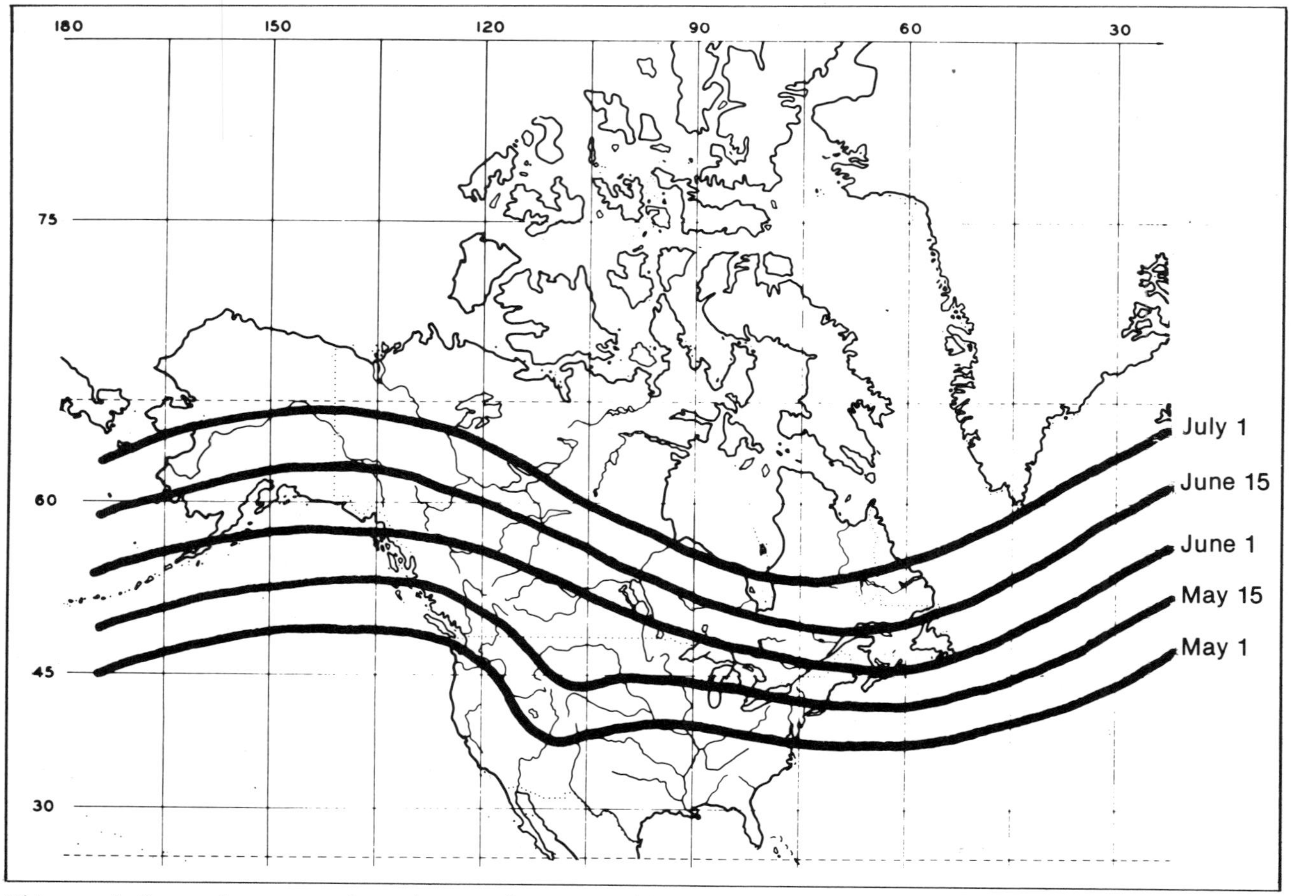


*This map indicates the average date when purple common lilac begins to bloom, based on temperature and solar radiation data from throughout the Northern Hemisphere.*

always account for local variations due to topography and other factors. However, Hopkins' rule of a delay of four days in the arrival of spring for each degree of latitude northward does closely match the old New England saying that spring advances at the rate of about 100 miles per week. In fact, phenological data based on observations of the first leaf and the first bloom of Persian lilacs during a five-year period from 1966 to 1970[4], showed an average northward advance in leaf and bloom dates in the Northeastern United States of 14 miles per day—almost exactly 100 miles each week! "Oldtimers" were well aware of nature's ways and paid close attention to her changes.

## Studying Indicator Plants

Although scattered phenological records existed during the late 1800s and early 1900s, there was no uniform program to collect and analyze the information for use by gardeners and farmers. The U. S. Weather Bureau attempted to establish a formal network of phenological stations shortly after the turn of the century, but the program collapsed because of inadequate financial support. Finally, during the latter part of the 1950s and the 1960s sufficient interest developed in phenological studies that a number of regional groups were set up to establish and maintain a network of phenological observation sites and a regular collection of phenological data, much as regular climatological data have been taken for decades.

One of the first such projects was set up at the Montana Agriculture Experiment Station at Bozeman, Montana in the spring of 1956, in cooperation with the U. S. Weather Bureau and local gardening interests. The following year the project was extended to cover eleven western states. Since then, at least one additional state has been added and regular observations of phenological events are made at more than two thousand sites using honeysuckle plants, and at more than one thousand locations where the common purple lilac is the indicator plant.* These two species (honeysuckle and purple lilac) were selected as excellent indicator plants because of their uniform response to a wide range of weather conditions, and their ability to thrive over wide geographical areas.[5] A similar phenological network, using a particular cultivar of Persian lilac and the same two cultivars of honeysuckle as used in the western network, was established in the North Central and Northeastern sections of the United States and into parts of Canada beginning in 1965.[6] A few sites were also set up in Alaska.

So phenological observations have come into their own in recent years, but are hardly a "household word." Yet in many ways, phenological information can be as useful to farmers and gardeners as the common measurements of temperature and precipitation. The growth and development of plants is strongly related to environmental conditions and represents the sum total of all the effects of air and soil temperatures, sunshine, precipitation, humidity, wind, dew, frost, etc. Thus, for example, comparing the time of first bloom of a certain variety of lilac with the time of the first bloom the previous year, indicates whether the current spring is as advanced as the past year's. At the same time, comparing the dates of first bloom of similar plants at two or more locations during the same season gives a good idea of the differences in climate between those areas. In practice, differences in phenological events are usually most strongly influenced by air temperature and sunshine, with moisture conditions playing a lesser role.

While lilacs and honeysuckle have been the

*Those plants used in phenological studies are commonly known as indicator plants, since they are used as "indicators" of general environmental conditions.

prime indicator plants used so far in the various network studies, they are by no means the only plants that can be used. Although the development of almost any perennial plant can be monitored each year, some plants do respond more directly to weather conditions than others. The following species of shrubs, for example, have been recommended as indicator plants for recording phenological observations in the Ohio River Basin and Lower Great Lakes Region[7]:

| Species | Season |
|---|---|
| Vernal witchazel *(Hamamelis vernalis)* | Late winter |
| Forsythia *(Forsythia intermedia)* | Early spring |
| Purple lilac *(Syringa vulgaris)* | Spring |
| Shrubby cinquefoil *(Potentilla fruficosa)* | Early summer |
| Glossy abelia *(Abelia grandiflora)* | Summer |
| Spiraea *(Spiraea bumalda)* | Mid-late summer |
| Witch hazel *(Hamamelis virginiana)* | Autumn |

The most useful event to observe in each species would probably be the date of first flower. Using the complete array of suggested plants would provide phenological data from late winter through autumn. There is no reason why these plants could not be used in other sections of the country where they can be grown successfully.

# USE IN AGRICULTURE

You may be asking, "What is the value of phenological observations to agriculture?" Basically, the phenological events can be used as a guide to other upcoming natural events and to various gardening practices such as planting and spraying. The same overall environmental conditions that determine, for instance, the time at which the first leaf buds break open on lilac plants also affect the growth

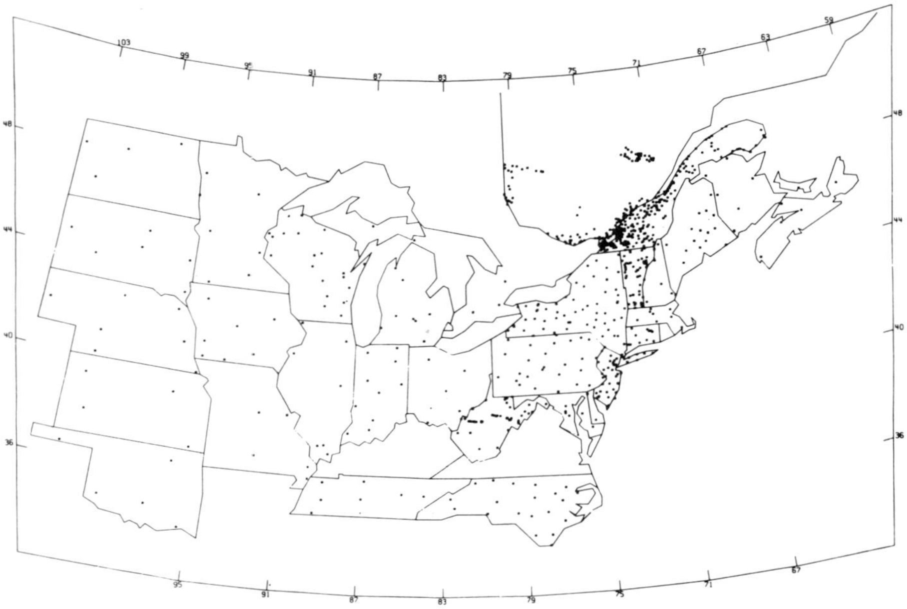

*The Northeast Lilac Phenology Network (NE-95) included 756 sites in 26 states in 1977.*

and development of other natural and cultivated species. Farmers in Montana, for example, have found that by making their first cutting of alfalfa within 10 days following the bloom of common lilac, they will usually complete their harvest before the alfalfa weevil eggs would probably be the date of first hatch. A delayed cutting increases the risk of weevil damage. The lilac bloom, the ripening of alfalfa, and the weevil egg hatch are all ultimately affected by the same general weather conditions.

Other examples of using phenological observations as guides to agricultural practices are occasionally found in agricultural publications, while some are simply passed on by word of mouth from one gardening generation to another. An old Indian custom led early New England farmers to plant corn when the leaves on elm trees were the "size of a squirrel's ear." Emergence of the European corn borer is reported to coincide closely with the blooming date of arrowwood.[8] If you don't have any elm trees nearby (and they become rarer each year) you can use oak leaves as a guide for planting corn. When the leaves become the size of a mouse's ear, corn can be safely planted—another custom passed on to early settlers by the Indians.

Quite a bit of work has been done relating the appearance of the cabbage maggot to various phenological events. On Long Island, growers are told to watch for the cabbage maggot adults about the time that the forsythia are blooming.[9] Injury to brassica* seedbeds by the cabbage maggot has been reduced in the North Carolina area by delaying planting until the flowering dogwoods are in full bloom.[10] And in Massachusetts, it has been found that the average date of the first appearance of the cabbage maggot eggs coincides closely with the full-pink stage of McIntosh apple blossom buds.[11]

*Brassica is the genus of plants which includes cabbage, cauliflower, and broccoli.

*Apple blossoms have been used as indicator plants in Massachusetts.*

Another phenological relationship was brought to my attention by Richard Hopp, professor emeritus at the University of Vermont, and a recognized authority on phenological observations. Based on his own records covering a period of a decade or more in northern New England, he suggests that cool season crops, such as peas, beets, and lettuce, may be safely planted shortly before or during the period when the first leaves are appearing on the common lilac. By the time the lilacs are in bloom, warm season species (corn, beans, etc.) may be sown with a minimum risk of late frost damage.

My own observations, taken in south-central Pennsylvania, while covering only a few growing seasons, suggest a similar relationship, although in at least one of those years a late season freeze occurred about a week after the first lilac bloom. In this particular case an unusual early-spring warm spell had hastened development of the lilac plants. This example simply serves as a reminder that the various relationships noted are based on long-term overall conditions, and do not always hold true each year.

Some growers set out tomato transplants when the first ladybugs are noted. However, it isn't clear whether ladybugs are used as a guide to help avoid frost damage or simply as a natural insect control.

Most older gardening practices, using events in the natural plant world as guidelines, were quite valid, and were generally based on many years of careful observation and experience. However, their usefulness for today's gardener has, unfortunately, been lessened for a number of reasons. For one thing, the use of such gardening aids as hot frames, black plastic mulch, mechanical frost protection devices, and specialized "starter" fertilizers have altered some of the natural relationships that existed in the past.

Secondly, and perhaps of even greater importance, has been the virtual explosion in new varieties of the more common garden plants in recent years. Many of these new varieties have been specifically bred to satisfy certain weather requirements—for example, lettuce that won't bolt so rapidly in southern climes and cantaloupes that can be successfully grown in northern regions. Thus, the old established relationships between natural plants and certain gardening practices no longer hold true in many cases. The basic *principles* behind these relationships still exist, however, and that is why a later part of this chapter discusses how growers today can establish new phenological relationships between plant species and garden crops grown in their region.

The use of phenological observations as a guide to agricultural operations is still in its infancy. Only in recent years have large-scale studies been started to investigate these relationships. Potential applications include using phenological data to estimate production conditions during the growing season, to help determine where a particular crop variety may prosper, and to aid in the timing of soil preparation, seeding, and spraying.

*First leaf of the lilac is one of the plant's key stages, and it may be used to help develop phenological relationships.*

Using phenological information in the selection of sites for new crop varieties is especially important at this time, when extensive efforts are underway to relieve chronic food shortages in various areas around the world. A new crop variety may show promise of large increases in productivity, but only in areas which have a favorable climate for its growth and harvest. Often, however, areas that are most in need of increased food production have poor climatic records on which to base decisions about plant adaptability. By studying and comparing data on natural vegetation, reliable estimates can be made about the climatic similarities between two or more locations, and a sound decision can then be made as to whether or not a particular variety should be introduced into a new region.

The potential for using phenological data as a guide to farming and gardening would seem to be almost unlimited. Much needs to be learned in this field and any prediction as to its ultimate usefulness can only be guessed at.

# ROLE OF THE GARDENER

Fine, you say—but how can the individual farmer and home gardener make use of phenological data? There are some old phenological relationships which have been used by gardeners for many years. Some were passed on to early settlers by Indians. However, many of these gardening guidelines are no longer as reliable as they once were. But, there is no reason why you, as an individual grower, cannot determine your own relationships between the development of natural vegetation and new garden plants—guides to gardening practices which will be of most value to your particular situation. This is one type of simple "research" you can perform on your own without waiting for crop scientists to do the work.

The first step in developing phenological relationships is to start keeping proper records. Keep them as regularly and as faithfully as any weather records you may keep. If you don't keep weather data then you can use the phenological observations as a substitute, although a combination of the two is the most useful.

What should you observe and record? The two most common and popular plants used so far in this country in phenological networks have been the common purple lilac and the honeysuckle. Other common perennial plants can also be used, depending on your location and climate. Probably you can use some plants already growing in your area, if not directly on your property. The main criteria for useful indicator plants are:

1) The species should be easily identifiable. Know what plants you are observing and, insofar as possible, observe the same plants year after year. Therefore, the plants should also be easily accessible for regular visits.
2) The plants should have easily observable stages of development, such as the breaking of leaf buds, flowering, and fruiting.
3) The plants you observe should be as representative of the surrounding environment as possible. That is, they should not have a peculiar micro-climate of their own because of nearness to a building, extreme slope, constant shade, or other distorting influences.
4) The plants should be free of serious injury and/or disease. An injured or diseased plant will not respond to weather conditions normally. One way to help avoid errors of this type is to select several plants of the same species and average the observations. For example, you might note the date of first flowers on three or four forsythia bushes and then determine the average date of first bloom for these plants.
5) The plants should have a succession of events which continues throughout much of the growing season. The longer into the summer you are able to make observations, the better you will be able to define the growing season and the more relationships you can develop.
6) The plants should be relatively common over large geographical areas. This requirement is not absolutely necessary, but it does make it more likely that information on their usefulness as indicator plants may be available from other studies.

*Common Purple Lilac*

## Events to Observe

The events to be observed depend partly on the plant involved. Certainly the time of the first opening of leaf buds in spring would be applicable to most deciduous plants. For species that flower, you could note the date of first flower, date of full bloom, and perhaps the date of last bloom. The date of the first appearance of fruit, where appropriate, would be a useful phenological observation.

The list of likely events for phenological study decreases somewhat as the growing season progresses, but even in the fall such items as the first indication of colored foliage, time of peak coloration, and date of leaf fall can be useful observations. And, while this discussion has focused on observations of events in the growth and development of plant life, various phenological observations may also be made of events in the animal kingdom, such as the migration times of birds and the hibernation dates of certain animals.

Phenological observations will tell you something about the current growing season, as compared to past years, but they can be much more valuable if you also maintain records of various events in your garden or cropland. After several years of collecting data you should begin to recognize relationships between certain phenological events and various gardening activities. As one example, the date of first flower on forsythia bushes may prove to be a good indicator of when to sow certain early season crops. To determine such a relationship, you will need records of the sowing dates for your crops as well as information showing any frost damage, the date on which the crop was ready for harvest, how good the yield was, both in quantity and quality, and any insect or disease problems you encountered.

*The purple lilac is a widely used indicator plant, with distinct stages of development. Observers keep track of the lilac in more than 25 states and five Canadian provinces.*

By keeping track of outbreaks of insects and/or disease you may well find a relationship between these outbreaks and certain phenological events, such as the cabbage-maggot/apple-blossom relationship discussed earlier in this chapter. The possibilities are almost limitless. The exciting thing about it is that here is one area where formal scientific study has been slow and the home gardener is in a good position to learn some useful information firsthand, before the "experts."

So add phenological records to whatever garden and weather information you already keep. You will find it fun and interesting, and worthwhile from a practical gardening standpoint.

# Part II:
# Weather and Crops

# 3
# Weather Effects on Crop Growth

Although most people recognize that weather exerts a strong influence on crop growth and development, the exact nature of weather/crop relationships is often not well known. And while it isn't necessary for the home gardener or farmer to learn all the detailed technical points of plant physiology, a general knowledge of how the weather affects crop growth can be useful as well as interesting to efficiency-minded growers. Hopefully, information presented in this chapter will help you better understand various suggestions in this book for using weather data to guide your gardening activities.

Two definitions are perhaps in order: The terms ***crop growth*** and ***crop development*** are often used interchangeably, although they refer to separate aspects of plant activity. Growth denotes an increase in the height, size, weight, or volume of a plant, while development refers to the progression of a plant through the stages of its life cycle. Thus, the growth of corn, for example, may refer to an increase in the stalk height or leaf size, while development alludes to the rate of progress from tasselling to silking, silking to ear formation, etc. For most purposes, growth and development can be used interchangeably, but in certain instances the distinction may be useful. Weather conditions that are favorable for growth of a particular plant, for example, may not be the most optimum for that plant's development, and vice versa.

All elements of the environment affect crop growth and development. In fact, the final product of a plant's life cycle is a result of the interaction of the various environmental elements; in many cases, it becomes almost impossible to separate the effects of any one factor from the rest. (This aspect of the response of plants to the various weather parameters makes some plants excellent "weather instruments.") Nevertheless, a general discussion of the individual importance of the major weather elements (temperature, moisture, and sunshine) will help to clarify their role in gardening and farming successes.

# AIR TEMPERATURE

One of the most important weather elements, and the one whose role in plant growth is perhaps best understood, is air temperature. Even the most casual backyard gardener recognizes the strong relationship between the air temperature and the growth of plants. It is tempting to simply state that as the temperature increases the growth and development of plants also increases. For most crops that is true—at least up to a point. But the real temperature/crop-growth relationship is not quite so simple, and a little knowledge of that relationship might enable you to time your plantings to get maximum production.

Nearly every physiological process involved in plant growth and development is strongly influenced by temperature. In general terms, temperature exerts its main influence by controlling the rate of the chemical reactions involved in the various growth processes within the plant. The solubility of minerals, the absorption by the plant of water, nutrients and gases, and the various diffusion processes that occur within the plant are all temperature dependent.

Air temperature also affects the hormone mechanisms involved in the fruiting and flowering of plants. In some species, low temperatures stimulate flowering, while other plants require relatively high temperatures before flowering is initiated. For example, the Boston variety of lettuce requires rather high temperatures for its flowers to develop. When temperatures are around 50° to 60° F. heads

*Weather Elements*

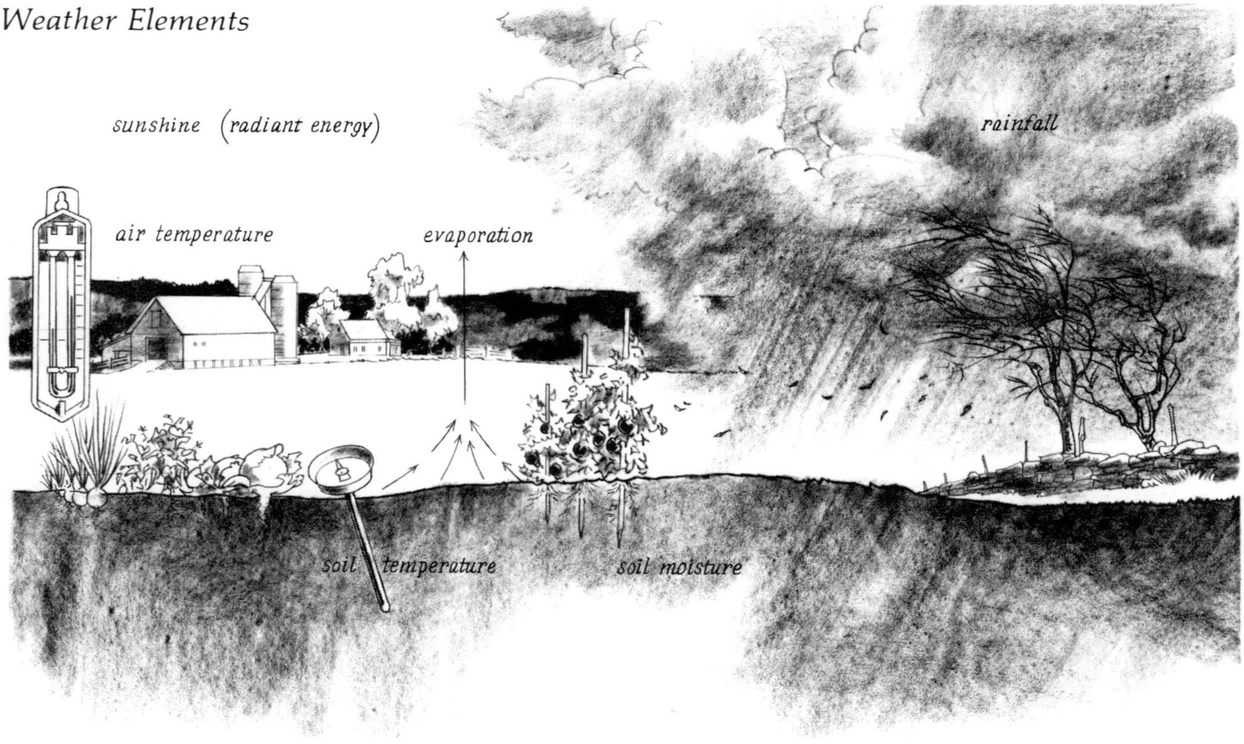

develop slowly and few, if any, flower stalks occur. At 70° to 80°, however, heads are unlikely to form while flower stalks develop readily. Thus, cool weather favors a good crop of lettuce, while warmer weather causes the plants to bolt and cease head production. On the other hand, celery, beets, and carrots are examples of crops which flower at relatively low temperatures. The rate at which plants ripen and mature is also strongly dependent on the air temperature.

The dormancy requirements of some plants are directly related to temperature. Peaches, for example, require a certain number of hours with air temperatures below about 45° in order to complete their annual development cycle and produce fruit the following season. Many biennial plants may fail to flower or set fruit the second summer (as they normally do) if they have not been exposed to the necessary low temperatures between the first and second growing seasons.

This requirement for a cold period may be satisfied by artificially induced low temperatures to speed up the development of the plants. Such a cold treatment is called ***vernalization***. One of the more common uses of this process is in the transformation of winter grains (such as wheat or rye) into spring varieties. By first exposing partially germinated seeds to a period of low (just above freezing) temperatures for about six weeks or longer, the seeds can be planted in the spring and complete their life cycle within one growing season, instead of two. In some cases, if the seeds are exposed to persistent high temperatures following their cold treatment, they may be devernalized and revert back to their natural life cycle. The duration of exposure to persistent high temperatures required for devernalization can vary considerably depending on the specific seed type and the degree of vernalization. One example would be onions, which are stored over winter at near freezing temperatures to prevent spoilage. In effect, they have been vernalized, and if planted directly from storage they would quickly flower and not form proper bulbs. Therefore, they are exposed to temperatures over 80° for several weeks to be devernalized. Usually, these seed treatments are done in temperature-controlled chambers.

## Temperatures and Crop Distribution

Most plant species survive within a general temperature range of about 32° to 122°, although some arctic species can withstand temperatures as low as minus 90°, while forms of algae have been found in hot springs near the boiling point of water (212°). Successful crop production usually occurs where growing season temperatures average between about 50° to

*Cool weather favors lettuce, while warm weather causes the plants to lengthen and bolt.*

105°. It is the *extremes* of temperature that determine whether or not a given species can survive in a particular region, while actual crop yields are more a function of the average temperatures over a growing season.

Generally speaking, temperature is the main factor in defining wide geographical "belts" around the globe in which various crops can be grown—regions such as the U.S. Corn Belt, the grain growing region of North America, or various areas of citrus fruit production. The distribution of grain crops, for example, is affected by such temperature related factors as the length of the growing season, the minimum winter temperatures (which affect survival), and the average and highest temperature shortly before harvest. Cool weather before grain harvest slows down the ripening process and delays maturity.

Temperature requirements for field corn have been determined probably better than have those for most other crops. One study[1] found that the normal temperature usually experienced during June in the U.S. Corn Belt is also the optimum temperature for corn growth and development during that period. The most favorable temperature during July and August, however, is about two degrees below the Corn Belt normal for those two months. Thus, somewhat cooler than normal weather in July and August favors increased yields. The greatest detrimental effect on corn yield comes from hot weather in August, during the time when the ears are filling. Heat hastens the loss of soil moisture, increases the respiration rate of the plants, and may produce a slight reduction in the rate of photosynthesis—the production of new plant material. The ideal daily maximum temperature for corn is about 86°, and when temperatures go over the 90° mark there is a noticeable negative effect on the corn growth. For every ten degrees of daily maximum temperature above 90° (for example, two days of 95° or one day of 100°) during July and August, there is a loss in yield of about one bushel per acre. Unusually low temperatures can also affect development of corn and reduce yields.

The actual effect of temperature on corn yields is partly dependent on moisture conditions. Hot weather during July and August will be less detrimental, for example, during a wet year than during a drought. Adequate soil moisture helps to reduce the stress placed on the crop due to the high temperature. This is true for all crops and is an important point for home gardeners to keep in mind.

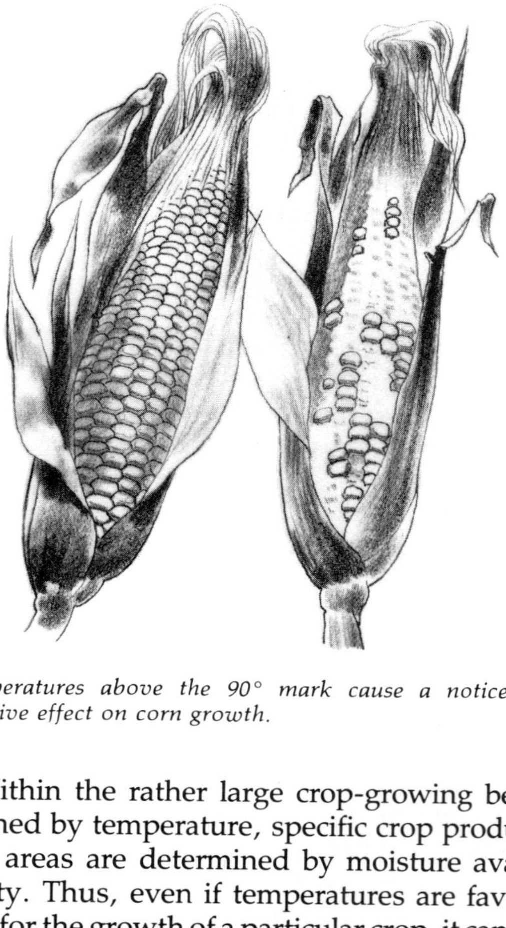

*Temperatures above the 90° mark cause a noticeable negative effect on corn growth.*

Within the rather large crop-growing belts defined by temperature, specific crop production areas are determined by moisture availability. Thus, even if temperatures are favorable for the growth of a particular crop, it can be successfully grown only if rainfall is sufficient or irrigation water is plentiful.

## Critical Temperatures

Each plant species has certain critical temperatures (sometimes called ***cardinal temperatures)*** which define the heat requirements needed for growth and development. These cardinal temperatures generally include the ***minimum*** (the lowest temperature at which growth will occur), the ***optimum*** (the temperature at which the rate of growth and development is the greatest), and the ***maximum*** (the highest temperature at which growth will occur). In addition, an upper and lower ***lethal*** temperature exist, at which death of the plant occurs. These lethal temperatures are more extreme than the minimum and maximum growth values.

Arctic plant species have optimum temperatures of about 50°, while temperate species respond most favorably to temperatures in the 77° to 86° range. Minimum growth temperatures for temperate-zone plants are about 40° while maximum growth values range between 95° and 104°. Sub-tropical plants, such as corn, have optimum temperatures around 86° to 95°, minima of about 50°, and maxima around 110° to 115°.

One of the difficulties in determining the "best" temperature range for a given crop is that each stage of development often has its own unique temperature requirements. The optimum temperature for seed germination is frequently lower than the optimum temperature for vegetative growth, which in turn may be different than the best value for flowering and fruiting. Thus, throughout the course of a plant's lifetime it may have two or three sets of critical temperature ranges.

Also, in some species the night temperatures are more important than daytime values, at least during certain stages of development. For example, the optimum nighttime temperature for tomatoes to set fruit is about 59° to 68°. Fruit set is much poorer at either higher or lower temperatures, regardless of how warm it is during the day. Blossoms begin to fall off when night temperatures remain above 72°. These optimum fruit-set temperatures, however, are lower than the optimum temperatures for general growth of the plant. Maximum yields occur when temperatures range between about 58° and 85°. Sugar beets grow best with night temperatures of about 68°, but at this temperature sugar production (which requires night temperatures at least 20° colder) is poor.

In addition, the temperature at one stage of development may influence the rate of development in a subsequent stage. The diurnal (daily) change in temperature is also important. Most plants grow and develop better when exposed to lower nighttime and higher daytime temperatures than when grown at constant temperatures day and night. High nighttime temperatures increase the plant's respiration rate which reduces the overall production of new plant material.

Specifying precise critical temperatures for a given crop is complicated by the development of so many new varieties of cultivated plants. Each variety has its own temperature requirements and some are bred specifically to adapt to either warmer or colder climates. Muskmelons, for example, have been developed which ripen satisfactorily in cooler northern climates where, at one time, they simply weren't grown. Various nut trees, such as pecan and almond, have also been developed for growth in northern regions. At the same time, newer varieties of lettuce have been bred to better withstand heat and are thus more suited for southern climes.

It is therefore difficult to specify exact critical

### Table 3: Cardinal Temperatures of Common Vegetables (°F.)

| Crop | Min. | Opt. | Max. |
|---|---|---|---|
| Cool Season Grains | 32-40 | 75-85 | 90-100 |
| Corn | 50 | 85-90 | 110-115 |
| Cucumbers | 48-50 | 75-80 | 95-105 |
| Melons | 59-65 | 86-98 | 110-120 |
| Peas | 38-42 | 50-60 | 70-75 |
| Potatoes (tubers) | 43-45 | 50-60 | 80-90 |
| Snap Beans | 50 | 80-90 | 100-110 |
| Tomatoes | 50-55 | 60-80 | 85-95 |
| (night temp. for fruit set) | 55-58 | 59-68 | 72 |

temperatures for a given species. At best, perhaps a range of temperatures rather than a precise value can be given. Table 3 lists some of the more commonly grown crop plants with approximate critical (cardinal) temperatures. These temperatures were compiled from various studies and reports; other sources may list slightly different values. To be on the safe side, you should check carefully the temperature requirements of the particular plant varieties you plan to grow. Often this information is available from the seed or plant supplier, but if not, check with your local extension service or county agent. In some cases, the exact temperature requirements for a particular variety may not be available. This presents an opportunity for home gardeners with their own weather station to determine the best temperature conditions. By keeping accurate temperature records and noting the growth characteristics of the crop and by comparing one season's yield with another's, you can see the effects of temperature variations.

# PLANT TEMPERATURES

While we generally talk about the effect of air temperature on plant growth, the actual element of greatest importance is the temperature of the plant tissue itself, especially the leaf temperature. Accurate measurement of leaf temperature, however, is quite difficult and expensive, so the air temperature is generally used as the best indicator of the heat energy available to the plant. Although it is unlikely that the average gardener or farmer will become involved in measuring plant temperatures, some general information may be of interest.

Leaves exposed to the sun are almost always warmer than the surrounding air, and they may be considerably warmer under certain conditions, as they absorb the radiant energy from the sun. Leaf temperatures are commonly five to ten degrees above air temperatures on sunny afternoons, and have been measured as much as 25° to 35° warmer than the air in some field experiments. Thus, on a hot summer afternoon with air temperatures in the 90s, plant surfaces may be as hot as 110° or more.

The greatest leaf-air temperature differences usually occur for brief periods at mid-day when leaves are exposed directly to bright summer sunshine and there is little or no air movement past the plant to help carry away the heat. Leaves which are partly dried by moisture stress are generally warmer than fully transpiring leaves. On cloudy days, or if the leaves are shaded, leaf temperatures will usually be within a few degrees of the air temperature.

At night, leaves will generally be at least several degrees cooler than the surrounding air, as they lose heat by radiation. This radiational heat loss from plant surfaces is greatest when skies are clear and winds are very light or calm. Leaf-air temperature differences of perhaps three to six degrees are most common, but differences as great as 10° to 15° have been reported. Perhaps now you can better understand why plants may exhibit frost damage even though the air temperature as recorded in a weather shelter remains above freezing. Despite the fact that the minimum shelter temperature may have been in the mid- or upper 30s, the leaves themselves may have cooled to as low as 28° to 30°—cold enough to produce frost injury.

If nighttime skies are cloudy, or the wind is strong, leaves will remain near the air temperature. Also, leaves lower down in the plant canopy and not directly exposed to the sky, will not cool much below the surrounding air.

*Although difficult to measure, it's interesting to note that leaf temperatures and air temperatures are often different.*

# SOIL TEMPERATURES

The importance of soil temperature to the growth and development of plants has been recognized for many years. In 1901, a committee of the American Association for the Advancement of Science concluded that soil temperatures, their daily and seasonal variations, and their vertical differences, "cannot fail to be of very great importance in the physical and chemical processes upon which growth, cell division, nutrition, and propogation depend."[2]

Soil temperatures have a critical influence on seed germination. Germination generally improves steadily with an increase in the warmth of the soil, although each plant variety has its own unique optimum temperature for emergence. For the direct seeding of tomatoes, for example, soil temperatures at the one- to two-inch depth should be in the 50° to 55° range. The minimum threshhold temperature* for planting cotton is about 68°. If cotton is planted at lower soil temperatures, early growth is extremely slow and the seedlings are susceptible to soil-borne diseases. Beans and corn will usually fail to germinate if the soil temperature at planting depth is below about 42° to 44°; temperatures closer to 50° are preferable. Minimum soil temperatures for the effective germination of some common crops are listed in Table 4.

In general, then, it usually doesn't pay to rush spring planting. Germination will either be very poor, or when the plants finally do emerge their early growth will be quite slow, predisposing the plants to various seedling diseases. A delay in planting until the soil has warmed sufficiently will mean more vigorous early growth and healthier seedlings. Carefully check the temperature requirements of the specific seeds you plan to use. If the seed packets do not clearly state the soil temperature requirements, check with your local Extension Service or agricultural college. An inexpensive soil thermometer is a valuable gardening aid.

Even when seedlings have emerged and air temperature becomes the primary heat factor, soil temperature remains important. With the plant's root system underground, soil temperature affects many important physical, chemical, and biological processes which influence the overall growth of the plant. For example, low soil temperatures reduce the rate of water absorption by plants, by increasing the resistance to water movement through the roots.[3] Thus the uptake of nutrients is also reduced by cool soils, with a resultant reduction in the rate of growth and development.

### Table 4: Minimum Soil Temperature for Germination

| Crop | Min Soil Temp. (°F.) |
|---|---|
| Cabbage | 38-40 |
| Carrots | 39-41 |
| Corn | 46-50 |
| Melons | 55-60 |
| Onions | 34-36 |
| Peas | 34-36 |
| Radish | 39-41 |
| Small Grains | 35-41 |
| Snap Beans | 48-50 |
| Sugar Beets | 39-41 |
| Tobacco | 55-57 |
| Tomatoes | 50-55 |

*Soil temperatures have a critical influence upon seed germination; these are minimums for effective germination of some common crops.*

Soil temperatures are also a controlling factor in the biological processes affecting the life cycle of soil micro-organisms. These tiny organisms, unseen by the naked eye, are important in the breakdown of organic matter within the soil and the availability of various nutrients to the plant. The process of nitrification within the soil (whereby nitrogen is made available to plants) does not begin until soil temperatures reach about 40° or higher, and proceeds most rapidly when the soil has warmed to around 80° to 90°. Soil temperatures are thus also a factor in the effectiveness of fertilizer applications. It has been found, for example, that liquid ammonia fertilization is most effective when soil temperatures are in the 50° to 55° range.

Thus, through its effect on these various underground processes, soil temperature influences the overall growth and development of the plant. Once again, exact temperature requirements depend on specific plant varieties and other soil conditions, especially moisture. Also, not all plants have been extensively investigated as to their soil temperature requirements. But, some studies have been made for a number of crop plants. The optimum soil temperature for the formation of potato tubers, for example, appears to be from about 60° to 70°. At lower temperatures there may actually be more tubers formed, but they are smaller than those formed in warmer soils. Once the soil reaches 80° to 85° the production of tubers essentially ceases. Top growth of potatoes is greatest with soil temperatures of about 70° to 72°, about the same temperature at which total yield reaches a maximum. The lowest soil temperature at which sprouts will develop is around 48° to 50°.

Soil temperatures should reach at least 50° to 55° before direct seeding of tomatoes is begun. Once the plants are established, warmer soils will produce greater top growth of the plants.

*High temperatures and low soil moisture produce a stress on the plants which reduces yields.*

*Beans produce their highest yields when soil temperatures are about 82° to 86°, assuming soil moisture is sufficient.*

At low soil temperatures root growth is extremely slow with little if any secondary root development. When soil temperatures reach about 6(  and higher, root growth significantly increases and there is profuse secondary root formation. This improved root growth is then reflected in greater top growth, and, eventually, improved yields. A more extensive root system also makes the plant more drought resistant, since the roots can extract moisture from more soil.

Studies have shown that field beans, including different varieties of common garden beans, produce their highest yield when soil temperatures are about 82° to 86°, assuming the soil moisture is sufficient. If the soil is dry, greatest yields occur at lower temperatures, around 68° to 72°. The higher temperatures with the lower soil moisture produce a stress on the plants which reduces yields. With soil temperatures below 68° a sharp reduction in yield can be expected.

Soil borne diseases are significantly affected by soil temperatures, with some disease organisms preferring warm soils, others requiring relatively cool soils. For example, Verticillium wilt of tomatoes prefers cool soils and develops best at about 68°, while Fusarium wilt is favored by warmer soils, with an optimum temperature of about 80° to 82°. An interesting example of the effect of soil temperature on these two fungus diseases occured in New Jersey early in the 1962 growing season.[4] Most of the commercially grown tomatoes in New Jersey had been bred for resistance to common Fusarium wilt, but were susceptible to Verticillium fungus which normally was found mainly in the cooler areas north and west of the state. However, a severe outbreak of the Verticillium wilt occurred in June 1962. Subsequent studies found that the southward movement of Verticillium fungus from its normal habitat was directly related to below normal soil temperatures that spring.

Nematodes, tiny roundworms found in soil and water, are strongly influenced by soil temperatures. The nematode that causes the disease known as root knot, for example, has an optimum soil temperature range of about 68° to 85°. Below about 55° to 60° there is little disease development, while extremely hot soil surface temperatures of around 105° or higher will destroy both larvae and eggs. In general, most nematodes do not multiply to a full population until the soil warms to about 70°.

One final note concerning soil temperatures. Those of you who record (or plan to record) soil temperatures in your garden will notice there is a lag in the time of the daily maximum and minimum soil temperatures when compared to the time of maximum and minimum air temperature. This lag increases with depth at roughly the rate of about one hour for each inch of depth. Thus, at the six-inch level the lowest temperature will usually occur around noon, while the daily maximum will occur around 8 or 9 p.m.

This lag is caused by the time required for heat to move through the soil, and varies with soil type and moisture content—wet soils warm more slowly than dry soils and also lose heat less rapidly. This same time lag occurs on an annual basis as well, with deeper levels reaching their lowest temperature late in the spring and their highest values well into fall. There is also a "damping" effect on temperatures with depth, both on a daily and annual basis. That is, the range between the highest and lowest temperatures at any given depth decreases as you go deeper into the soil. For example, the difference between the daily maximum and minimum soil temperature at the one-inch level on a particular day might be 16°, at four inches it may be only 8° and at eight inches perhaps only 4°. Eventually, at some depth (the exact level varies considerably with soil type, soil moisture and other factors, but about 20 inches can be used as a general average), there is no daily cycle in the soil temperature, only a slow seasonal change. Even the seasonal changes are damped-out eventually, although at depths which are of no real concern to farmers or gardeners.

## RADIATION

The ultimate source of all energy for plant growth is the sun, but how many of us stop to consider exactly what occurs when sunlight strikes a living plant?

Sunlight, or any light for that matter, is energy transmitted in a form we call radiation. In simple terms, we can envision radiation as being waves of energy, similar to waves of water in the ocean. These waves vary in their ***wavelength***, which, as shown in the figure (***Wave Form of Radiant Energy***) is the distance between wave crests, or troughs.

WAVE FORM OF RADIANT ENERGY

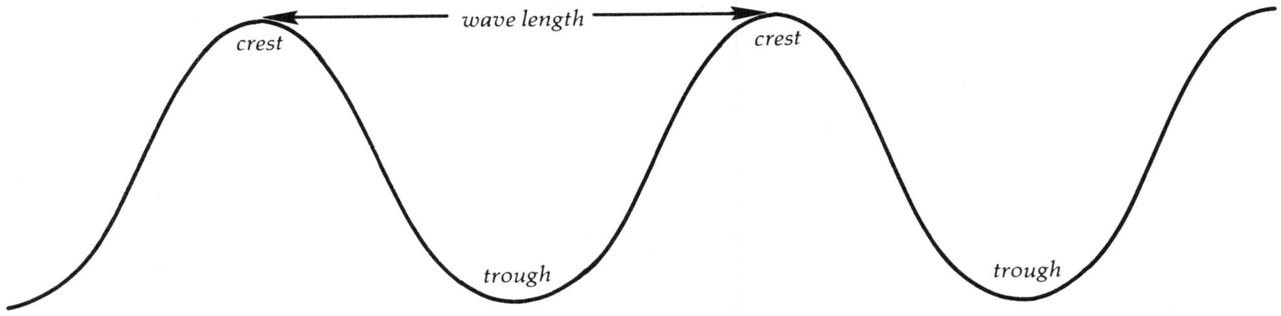


Wavelengths are measured in various units, one of the most common being the micron. One micron is 0.000039 inches in length—quite a short distance! Put another way, there are approximately twenty-five thousand microns in one inch. There is a wide range of radiation wavelengths, from extremely shortwave cosmic rays to the very long radio waves. The complete range of wavelengths is often called the radiation spectrum. As you can see from the figure (***The Radiation Spectrum***), the visible part of the spectrum is quite small, ranging from about 0.36 to 0.76 microns. Within this visible spectrum (the only part we can "see" with our eyes) differences in wavelength produce what we call colors (see ***The Visible Spectrum***). Thus, what we refer to as sunshine is in reality only a portion of all the radiation emitted by the sun and entering the earth's atmosphere. Most of the radiation striking our planet originates at the sun (referred to as solar radiation), but some (especially the high-energy cosmic rays) have their origin in deep space. The earth's atmosphere blocks out much of the shortwave radiation (fortunately for us) and there is not much natural radiation occurring at the extremely long wavelengths. Therefore, it turns out that about 46 per cent of all the radiation actually reaching the earth's surface is in the visible part of the spectrum. This is what we call "sunlight."

THE RADIATION SPECTRUM

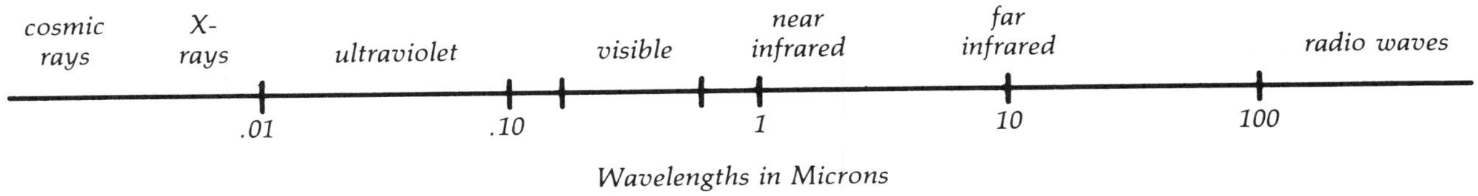


THE VISIBLE SPECTRUM

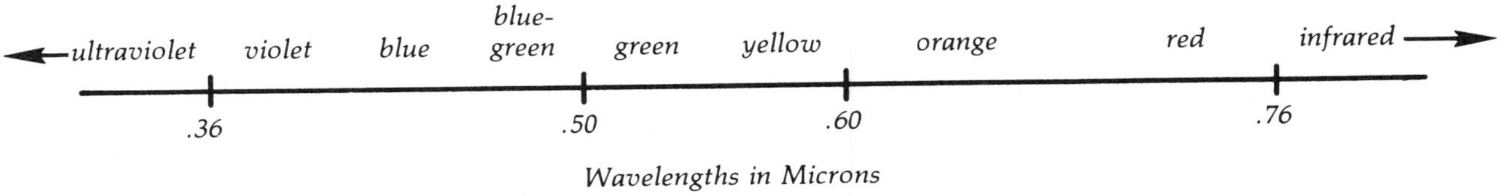

may reach critically high levels, causing further damage.

Only a very small percentage (perhaps no more than one or two per cent) of the solar energy absorbed by plants is actually used in the process we call photosynthesis—the ultimate beneficial use of this energy by plants. In photosynthesis, sunlight, water and carbon dioxide combine to produce new plant material. Without sunlight, the basic photosynthetic reaction could not occur and plant life as we know it would cease. In general terms, the photosynthetic reaction is:

**carbon dioxide + water + sunlight = organic molecules + oxygen**

This photosynthetic activity is centered in tiny microscopic bodies called chloroplasts, which are found inside individual plant cells. A substance called chlorophyl within these chloroplasts is capable of absorbing certain wavelengths of solar energy, using this energy to split water molecules, and forcing a combination with carbon dioxide ($CO_2$) to form carbohydrate (organic) molecules—the building blocks of plant material. Thus, the sun's energy, combined with water and carbon dioxide, is transformed into new plant material and

## Use of Radiation By Plants

It is that portion of the solar radiation that is absorbed by plant and soil surfaces which is our chief interest in this book. This absorbed energy is used in several ways. First of all, if there is moisture on the plant or soil surface some of the radiant energy is used to evaporate the water. Energy used in evaporation is not available for producing heat. That is one of the basic reasons why wet soils warm up less rapidly than drier soils. (Another reason is the greater heat capacity of wet soils as compared to dry ones.) Additional energy is used in the transpiration of water by plants (the loss of moisture from within a plant through its leaves). Various studies have found that in tropical regions, and in middle latitudes during the summer, as much as 75 to 85 per cent of the radiation absorbed by a crop canopy may be used in evaporating and transpiring water—if the soil moisture is adequate.

If there is little moisture to evaporate, most of the solar radiation will be converted into heat, causing high surface temperatures. Thus, as crops become dry, not only are they affected by moisture stress, but the plant temperatures

*Photosynthesis uses light energy to convert carbon dioxide ($CO_2$) and water ($H_2O$) into organic compounds and oxygen ($O_2$).*

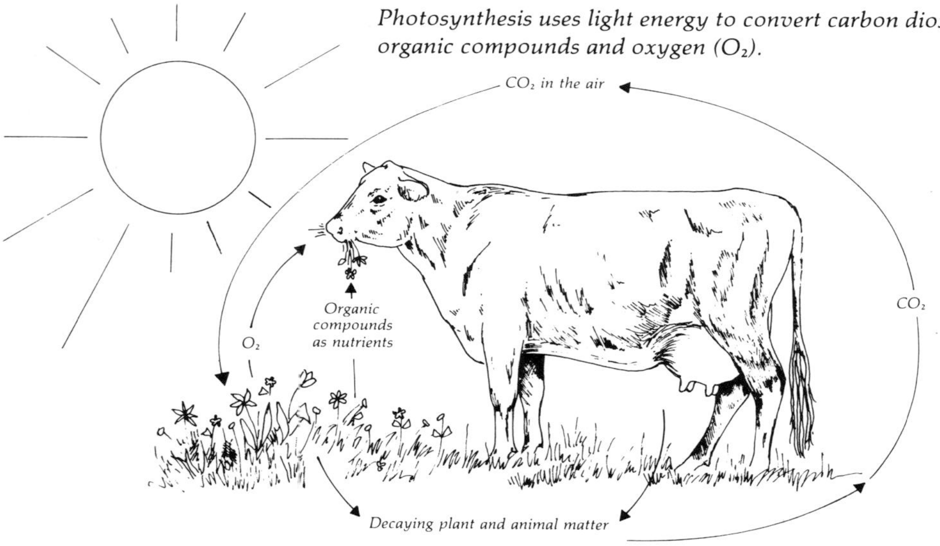

is the ultimate "driving force" behind the growth and development of plant life.

Within the visible and near-visible spectrum, the various wavelengths make differing contributions to plant growth and development. Radiation in the blue and red portions of the spectrum appears to be the most important for photosynthetic activity, while chlorophyl development is most dependent on radiant energy associated with blue and orange/red light. Phototropism (the effect of light on the direction of growth of a plant) is mainly affected by radiation in the near-ultraviolet to blue/green portion of the spectrum. Photoperiodism (the effect of daylength on plant development) appears to be most sensitive to red light. Plants absorb primarily in the blue and red portions of the spectrum and only slightly in the green. Much of the green light is thus reflected or transmitted on through the plant, producing the green color normally associated with living vegetation.

In general, if water and nutrients are not a limiting factor—that is, if the soil moisture and fertility are sufficient—the productivity of a given plant or crop depends primarily on the radiation and temperature. Under these conditions, radiation then becomes a limiting factor. Put another way, if the radiant energy available to the crop could be increased, productivity would also increase. One study[5], for example, found that by placing aluminum foil alongside the outer row of a corn field, and reflecting an increased amount of sunlight onto the plants, the yield of that outside row of corn was increased substantially. Experiments of this nature would be quite interesting in home gardens. Keep in mind, though, that additional radiation will only increase yields if water and nutrient levels are sufficient to support the additional growth. Also, additional light will mainly affect those crop canopies that are heavily shaded, since individual plants, and open plant canopies with little shade, appear to reach their maximum growth capacity at light levels well below full sunlight.

## Energy Efficient Plants

Photosynthetic efficiencies (measured by the amount of solar energy required to produce a given amount of plant material) are generally not very high for most plant species. Large variations have been reported in studies attempting to determine the energy efficiency of various plants. Much of this variation has to do with differences in experimental conditions and the difficulty of making some of the necessary measurements. Generally, the most efficient species (those that make the best use of sunlight) include corn, sorghum, and sunflowers. These plants may reach maximum photosynthetic efficiencies of about four to six per cent, although over a full growing season the average efficiency is generally less. On a worldwide basis, the efficiency of crop plants is estimated to be about two to two and a half per cent. Increasing the efficiency with which crops utilize sunlight appears to be one of the more promising means to significantly increase world food productions. Breeding programs are underway to develop new plant varieties that make more efficient use of sunlight.

For a given crop, the greatest photosynthetic efficiency is usually achieved with a closed canopy. Under such conditions, with closely spaced crops, nearly all of the solar energy strikes the plants rather than the soil surface. Open canopies cause low energy efficiencies, as do moisture stress on the plants, lack of sufficient nutrients, and low carbon dioxide. Thus, closely spaced crop rows, which produce a closed canopy quickly and permit maximum absorption of solar radiation, may not be the most efficient planting scheme if moisture and nutrient deficiencies exist. If, however, you can provide enough water and nutrients to your garden, then closely spaced rows will allow you to produce maximum yields from available space.

Shading reduces crop yields, primarily because of the loss of radiant energy to the plants. Even under "open" trees, for example, the average light intensity is only about five to ten per cent of full sunlight. In addition to reducing the radiant energy available for photosynthesis, shading also may result in the partial closure of tiny openings (called stomates) on the leaf surfaces, restricting the uptake of carbon dioxide and further reducing the rate of photosynthesis. One study of the effects of shade on corn found that not only did strong shading significantly reduce the overall yield, but it also increased the number of barren and partly barren ears.

## Table 5: Daylength Requirements of Plants

| Short-Day Plants | Long-Day Plants | Indeterminant Plants |
|---|---|---|
| aster | barley | asparagus |
| chrysanthemum | beets | azalea |
| cocklebur | clover | begonia |
| cosmos | dill | buckwheat |
| morning glory | goldenrod | carrot |
| orchid | grains (most) | celery |
| pointsettia | hibiscus | chickweed |
| ragweed | lettuce | corn |
| salvia | petunia | cotton |
| soybean | potato | cyclamen |
| strawberry | radish | dandelion |
| sweet potato | ryegrass | english holly |
| violet | spinach | geranium |
| | timothy | narcissus |
| | | primrose |
| | | tobacco (most) |
| | | tomato |
| | | zinnia |

*Short-day plants bloom or fruit under natural fall or spring conditions in mid-latitude regions, while long-day varieties flower during the summer. Indeterminant plants appear to be unaffected by daylength.*

## Photoperiodism

For some plants, the primary factor in determining the time of flowering or fruiting is the daylength. Temperature, moisture, and nutrient conditions must, of course, be in a generally favorable range, but the final determinant is daylength. This effect of daylength on plants is known as *photoperiodism*. In reality, it is the length of the *dark*, rather than the light, period that affects the plant's development. Thus, even a brief period of low-intensity light during a dark period will negate the effect of a long night. On the other hand, a brief period of darkness during a light period will have little or no effect. That is why, for example, if a pointsettia, a short-day (long-night) plant, is exposed to light for even a brief time during its required dark period, its flowering will be affected.

The primary effect of daylength is usually on a plant's reproductive growth—that is, its flowering and/or fruiting. On the basis of this effect, plants are popularly characterized as short-day or long-day plants, even though it would be more accurate to speak of long-night or short-night species. A plant pigment called phytochrome appears to play a major role in determining a plant's response to daylength. Short-day (long-night) plants bloom, or fruit, under conditions naturally found during spring or fall in mid-latitude regions, while long-day (short-night) varieties flower during the period of long summer days and short nights. In the field, control of daylength is difficult, although brief light interruptions at night are used by some sugar cane growers in Hawaii to control the time of flowering. Greenhouse conditions, however, can be adjusted to produce desired flowering or fruiting at various times of the year. Table 5 lists examples of long- and short-day plants, as well as some which appear to be unaffected by daylength, known as indeterminant species.

*Strawberries, called* **short-day plants**, *are among the plants that bloom and bear fruit in the spring when days are still relatively short and nights are long.*

Now, perhaps, you can better understand why strawberries (short-day plants) bloom and bear fruit in spring, while days are still rather short. Short-day flowers (such as the violet) also blossom at this time of the year, or later in the season (ragweed or chrysanthemums, for example) as the day-length begins to shorten again and the nights become longer. Long-day vegetables, such as spinach, radishes, and lettuce, remain in their vegetative (and edible) stage during the spring and again in the fall, but begin to bolt (set flowers) as the warmer, longer days of summer approach. Through careful breeding programs, new varieties of certain plants have been produced with altered daylength requirements.

## Radiation Measurements

Solar radiation is generally measured in terms of the number of calories (a measure of energy) striking a surface area, usually a square centimeter, or $cm^2$. The resulting unit, calories per square centimeter ($cal/cm^2$) is often called a langley (ly). Thus,

$$\textbf{1 langley} = \textbf{1 cal/cm}^2$$

Instantaneous readings of solar radiation are usually given in terms of langleys per minute (ly/min)—the amount of energy striking a square centimeter each minute. (One ly/min is equivalent to about 200 BTU per square foot per hour.) Daily and seasonal totals of radiant energy may be accumulated as a measure of the total energy available to plants for growth and development. Instruments are available to measure and record solar radiation. The amount of plant growth and stage of development of a crop at any given time during the growing season depends on the accumulated radiant energy.

The maximum radiation reaching the *top* of the earth's atmosphere from the sun is just under 2 ly/min. This figure is known as the ***solar constant.*** The value of the solar constant apparently varies very little, if at all, over long periods of time. The amount of radiation reaching the earth's surface, however, is much less than the value at the top of the atmosphere and varies considerably, depending on astronomical factors such as season of the year and time of day, and on atmospheric conditions including cloud cover, pollution, and moisture content.

The actual energy received at the surface shows a strong diurnal (daily) and annual variation. Daily values rise from zero just before daybreak to a maximum at solar noon and return to zero after evening twilight. In temperate and polar latitudes, annual values reach a maximum at the time of the summer solstice (about June 21 in the northern hemisphere) and a minimum at the winter solstice (about December 21). Equatorial regions show relatively little annual variation, with minor peaks at the spring and fall equinoxes (March 21 and September 21, respectively), and weak minima at the solstices. Maximum daily values of solar radiation run about 500 to 600 ly in temperate regions during mid-summer.

*Radiant energy determines the amount of plant growth and stage of development of a crop at any given time during the growing season.*

# MOISTURE

Probably the most important element associated with plant life is water. Water constitutes much of the bulk of all plant material, ranging from a few percent in dried seeds to about 40 per cent in dormant woody tissue, and up to about 90 to 95 per cent in succulent fruits such as watermelon. On the average, most actively growing plants consist of about 75 to 90 per cent water. Thus, plants generally contain more water than all other materials combined.

What role does water play in plant growth? Within the plant, water acts as a solvent and carries nutrients absorbed from the soil through the root system and into various parts of the plant. Water also preserves the turgor of individual cells—that is, it provides the pressure that maintains the cells' shape and rigidity. In addition, water is a nutrient required for the production of new compounds within plants. It is a vital ingredient in the photosynthetic process. Loss of turgor within the plant, due to decreased water content, results in the closing of stomates, a reduction in the rate of photosynthesis, disruption of carbohydrate and nitrogen metabolism, and a slowing down or complete cessation of cell division and enlargement. In other words, overall plant growth and development is slowed or halted.

The amount of water used by growing crops can be considerable. Studies in the U.S. Corn Belt have shown that up to about 15 inches or more of water may be lost by soil evaporation and transpiration from corn plants during the growing season. On a hot, dry day the crop may lose more than its own weight of water. The loss of water from deciduous forests in the humid southeastern United States has been estimated at 15 to 25 inches per year. The approximate daily water use of several popular crop plants is listed in Table 6. These values are general and will normally vary with local soil and climate conditions, as well as among different varieties of a given species.

**Table 6: Water Usage of Selected Crops**

(Average Peak Use, Inches Per Day)

| | TEMPERATURE AND MOISTURE CONDITIONS | |
|---|---|---|
| Crop | Cool/Humid | Hot/Dry |
| General Vegetables | 0.12 | 0.25 |
| Potatoes | 0.10 | 0.25 |
| Tomatoes | 0.14 | 0.27 |
| Fruit Trees (Deciduous) | 0.15 | 0.35 |
| Citrus Trees | 0.10 | 0.25 |

Another way to describe water usage is in terms of the weight of water that is absorbed for each unit of dry plant matter produced. Most crop plants have a water/dry-matter ratio ranging from several hundred to nearly one thousand, with some leafy vegetables having a ratio of two thousand or more. A ratio of five hundred, for example, means that five hundred pounds of water are required by the plant to manufacture one pound of plant material (500/1).

These values may seem high. Most of the water absorbed from the soil, however, is transpired into the atmosphere and takes no permanent part in plant growth processes. Less than one per cent of the water absorbed is used in photosynthesis. Most of the water retained by the plant becomes physically and chemically a part of the plant material.

One way to reduce the effects of drought is to breed plants which are more efficient water users and more drought-resistant. For example, new semi-dwarf strains of wheat have been developed that are able to produce more grain for each pound of water used than are the taller wheat varieties.

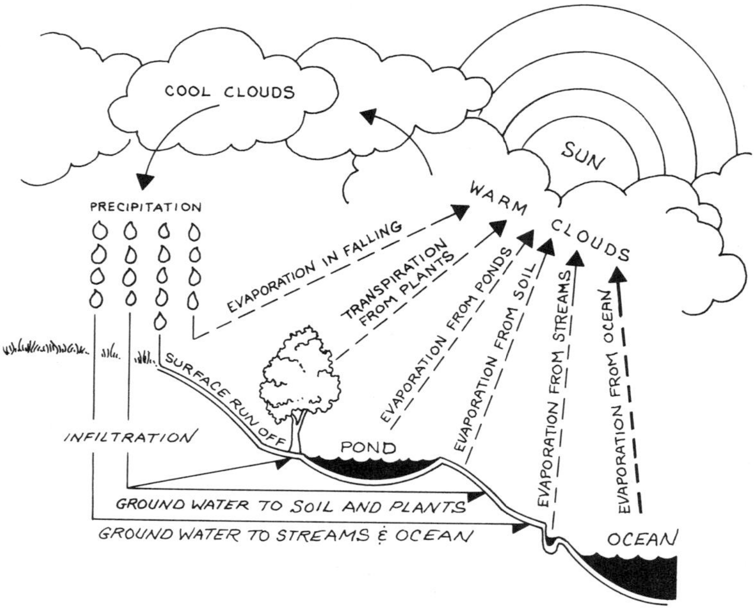


*Hydrologic Cycle. The sun evaporates water from the sea and the land to form clouds; then rain falls. Part of the rainfall is intercepted by plants; part is used immediately for plant growth: part is held by the soil as reserve for later use; part goes to underground resevoirs to sustain wells, springs and streams; and part runs off to form rivers, which return to the sea.*

## Water Loss

In a garden or field, moisture is lost to the atmosphere in one of two ways. First, free water is evaporated from the soil and plant surfaces. This is usually referred to simply as evaporation loss. Secondly, water is lost through a process known as ***transpiration.*** Most of the moisture taken up by plants is lost to the atmosphere. This water is evaporated within the plant tissue and the resulting water vapor escapes to the atmosphere through tiny pore-like openings on the plant surface, known as ***stomates***. In most plants, the principal location of these stomates is the leaves. This loss of water vapor through the plant surface is called ***transpiration.***

The number and size of stomates varies among plant species. They may occur on both top and bottom leaf surfaces or only on one surface. Bean plants, for example, have approximately four thousand stomates per square centimeter ($cm^2$)* on the tops of their leaves, and about twenty-eight thousand per $cm^2$ on their leaf bottoms. Corn has almost five thousand five hundred stomates per $cm^2$ on its leaf tops and about seven thousand per $cm^2$ on the bottoms. Tomato leaves contain around one thousand two hundred stomates per $cm^2$ on the top and thirteen thousand per $cm^2$ on the bottom. Apple leaves generally have their stomates only on the bottom side, with approximately thirty thousand per $cm^2$. So the stomates are miscroscopic in size but, because of their great numbers, permit large volumes of water vapor to escape from the plants.

The rate of water loss through the plant pores depends on the size of their openings, which varies according to environmental conditions. In general, stomates will open with exposure to light and close in the dark. Water loss by transpiration is therefore generally negligible at night for most species. At temperatures near freezing, stomates usually remain closed, even during the daytime, opening gradually as the temperature increases. With extremely high temperatures (about 100° and higher) some stomates may open even in the dark. The size of the stomatal openings is also

*To convert to number of stomates per square inch, multiply the values given by 6.25.

dependent on the overall moisture conditions within the plant. If an internal water stress develops, the pores will gradually close, to reduce the amount of water loss and preserve the remaining moisture. The actual amount of water lost through transpiration therefore depends on weather and environmental conditions. The main factors influencing the transpiration rate include:

1) Solar radiation—an important control factor in the opening and closing of stomates; it provides the energy to evaporate water within the plant.
2) Humidity—the higher the relative humidity of the atmosphere, the less the amount of transpiration that will occur. Drier air can hold more additional moisture than can moist air.
3) Temperature—warmer temperatures generally produce higher rates of transpiration, since warm air can hold more moisture than can cold air.
4) Wind—generally an increase in wind speed will increase the rate of transpiration by removing the water vapor next to the leaf surface and replacing it with drier air which can then absorb additional moisture.
5) Soil moisture—as the soil moisture decreases and a moisture deficit develops within the plant, transpiration is reduced, due primarily to closing of the stomates.

Different species of plants will transpire at different rates under the same atmospheric conditions.

The total amount of evaporation and transpiration (usually referred to collectively as evapotranspiration, or ET) varies daily, as shown in the figure *(Daily Evapotranspiration Cycle)*. The two-part illustration shows the usual variation in evapotranspiration when soil moisture is adequate and it shows variations that occur when soil moisture is reduced. In the

DAILY
EVAPOTRANSPIRATION CYCLE

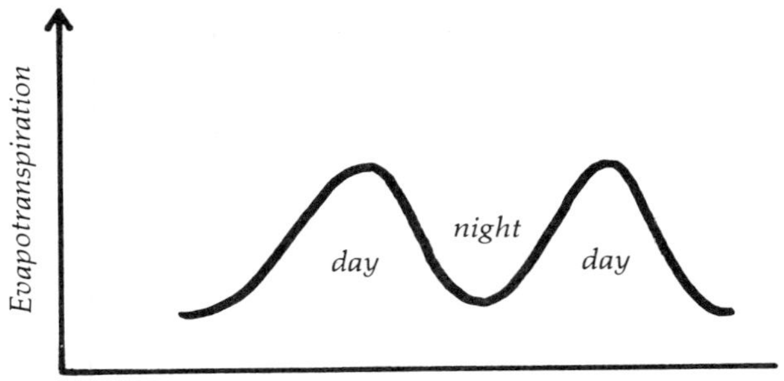


*ET Cycle With Plentiful Soil Moisture*

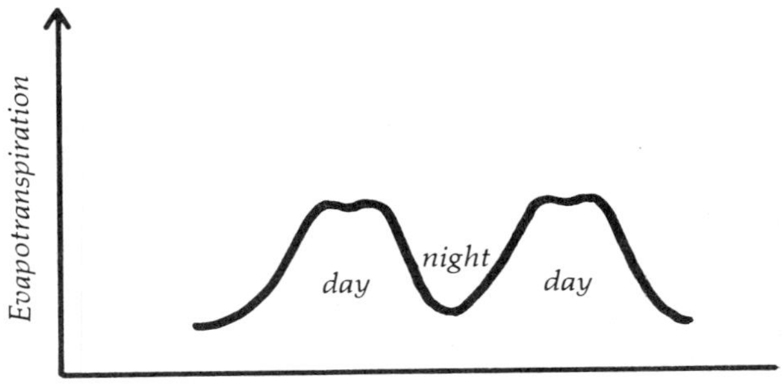


*ET Cycle During Moisture Stress*

latter case, during the warmest part of the day, the demand for water by the plant, to replace that lost to the atmosphere, exceeds the amount that the plant can absorb, resulting in an internal moisture stress. This stress, in turn, causes the stomates to close, reducing the transpiration rate. As the temperature and solar radiation decrease toward evening, the stress is reduced, stomates reopen, and a more normal transpiration rate returns. This increased water loss during the warmest part of the day can lead to temporary wilting if the difference between the transpiration rate and the rate of water absorption by the root system is great enough. In general, the overall water balance within the plant depends on the relative rates of water loss through transpiration and water gain through absorption by the root system. This points out the importance of having a healthy, well-developed root system (see Chapter 5) and maintaining sufficient moisture in the soil to meet the crop's peak demand on hot, dry days.

When the evaporative demand for water is extremely high, a stress may develop within the plant even though the soil moisture is adequate, simply because the plant's root system is unable to absorb water fast enough to replace that lost through transpiration.

## Effects of Drought

Crop yields in the United States are more often reduced by a lack of water than by any other single factor. Freeze, hail, disease—all of these hazards can and do seriously affect our crops. None of them, however, is as serious, year-in and year-out, as drought. Even in years with generally sufficient rainfall over a given region, there often are local areas of water shortage due to the uneven nature of summertime showers.

There are three basic types of plants which can resist drought, or at least minimize its effects. The "ephemerals" are best represented by desert flowers—plants which produce seeds that can lie dormant in dry soil for years, if necessary, waiting for a brief shot of moisture. When the rains come, the seeds germinate very quickly and the life cycle of the plant is completed within a few weeks. A second group of drought resistant plants are the succulents, which include cacti. These species accumulate large amounts of water within fleshy stems and leaves, and can survive over long periods of drought.

The third group of drought tolerant plants is perhaps the only truly drought *resistant* type, in that these plants can actually withstand severe reductions in their internal water content without serious damage. Their cells can, in some manner, endure dessication without permanent injury.

Unfortunately, most garden plants and field crops are not members of any of these drought tolerant groups. Thus, they require an adequate supply of water for growth and productivity, and, for many crops, the most critical time is the period of flowering and/or fruiting. Snap beans are most susceptible to drought-caused yield losses during the time of flowering and pod formation. Corn is extremely susceptible to dry periods during the silking and tasselling period, and on through the time of ear filling. One study of field corn found that if the soil moisture dropped to the wilting point for even one or two days near tasselling, there would be a yield reduction of up to 25 per cent. A brief summary of the most moisture-sensitive periods for various plants is given in Table 7.

### Table 7: Critical Moisture-Sensitive Periods for Selected Crops

| Crop | Critical Period |
|---|---|
| Beans | flowering and pod development |
| Broccoli | head formation and enlargement |
| Cabbage | head formation and enlargement |
| Corn | tasselling through silking; ear filling |
| Oats | beginning of ear emergence |
| Onions | bulb formation |
| Peas | flowering and pod filling |
| Potatoes | after initial tuber formation |
| Tomatoes | blossom set through fruit enlargement |
| Wheat | heading and filling |

If plants are subjected to moisture stress early in their development, plant size and vigor are reduced. If the stress comes later, during the critical flowering or fruit-forming stages, the results are usually more serious. In most cases, if the moisture stress has been slight (as evidenced, for example, by temporary wilting during mid-day) the addition of sufficient water to the soil will bring almost immediate improvement and little yield loss. If the stress is more severe, recovery will take up to several days after water has been applied, and a reduction in yield is likely. Even when the physcial appearance of the plants indicates full recovery, the effect of the stress on the rate of photosynthesis will usually linger.

## Soil Moisture

In discussing soil moisture, two terms are frequently used which sometimes cause confusion. The maximum amount of water that can be held by the soil after suface water has been drained off and water lost by gravity is removed, is referred to as the ***field capacity***, (FC). The percentage of moisture held by soils at the field capacity varies from as little as five per cent in very sandy soils to about 45 per cent in clays. Most soils return to field capacity within one to three days following significant rainfall or irrigation.

The moisture content of the soil below which irreversible wilting of vegetation occurs if known as the ***permanent wilting point*** (PWP). The percentage of moisture remaining in the soil at the permanent wilting point varies from about three per cent in sand to about 30 per cent in clayey soils. The ***available water*** is that moisture held by the soil between the field capacity and the permanent wilting point. It represents the water that can be used by plants.

The average amount of available water in the soil (in inches of water per foot of soil depth) varies from about 0.5 inch in very course sands, to around 2.0 to 3.0 inches in fine-textured clays, and clay loams. For most plants, the best moisture conditions exists when the soil is at or just below field capacity. Above field capacity, the excess water reduces the aeration of the soil to the point where growth is retarded. The accompanying lack of oxygen reduces the absorption of both water and mineral nutrients by the plant roots. As the amount of moisture decreases below the field capacity, due to evapotranspiration, there is an increasing risk of moisture stress in the plants. Initially, temporary wilting occurs during mid-day. As the soil moisture drops below the permanent wilting point, growth ceases and death eventually occurs.

For those interested in general soil moisture conditions over large regions (for example, the Corn Belt), a Crop Moisture Index (CMI) was

### Water-Holding Characteristics of Various Soil Textural Classes

| | Approximate Inches of Water per Foot of Soil Depth | | |
|---|---|---|---|
| Soil Textural Class | Field Capacity | Wilting Point | Available Water Capacity |
| *Sand* | *1.2* | *0.3* | *0.9* |
| *Fine sand* | *1.5* | *0.4* | *1.1* |
| *Sandy loam* | *1.9* | *0.6* | *1.3* |
| *Fine sandy loam* | *2.5* | *0.8* | *1.7* |
| *Loam* | *3.2* | *1.2* | *2.0* |
| *Silt loam* | *3.5* | *1.4* | *2.1* |
| *Sandy clay loam* | *3.7* | *1.6* | *2.1* |
| *Clay loam* | *3.8* | *1.8* | *2.0* |
| *Silty clay loam* | *3.8* | *2.1* | *1.7* |
| *Clay* | *3.9* | *2.4* | *1.5* |

*For most plants, the best moisture condition exists when soil moisture is at or just below field capacity. The amount of available water that a soil will hold is very important when irrigating because it governs the amount of water that can be effectively applied at each irrigation and it influences the time interval between irrigations.*

developed several years ago by Wayne Palmer of the U.S. Weather Bureau.[6] Maps of CMI values for the United States are published during the growing season in the *Weekly Weather and Crop Bulletin* (see Appendix C).

## CARBON DIOXIDE

One other environmental factor plays a crucial role in plant growth. That element is carbon dioxide ($CO_2$). Although it is not given much thought by farmers or gardeners, nor is it widely measured and reported, carbon dioxide is one of the prime building blocks of plant tissue. Carbon dioxide combines with water in the presence of sunlight to form carbohydrates and oxygen during the photosynthetic process. During respiration, the process is reversed and $CO_2$ is released.

Carbon dioxide is always present in the atmosphere and is fairly uniformly distributed, with a concentration of about 330 parts per million (ppm), or about 0.033 per cent of the total air. It is thus a minor gas in terms of its total concentration, but major in its effect on plant and animal life, as well as our climate. The primary sources of $CO_2$ include respiration by plants and animals, volcanic eruptions, and the combustion of organic materials such as wood, coal, oil, and gasoline—the "fossil fuels."

The world-wide concentration of carbon dioxide has been gradually increasing in recent decades, mainly due to increased burning of fossil fuels. Most estimates of atmospheric $CO_2$ content during the 1800s range from around 280 to 290 ppm, so today's figure of 330 ppm represents an increase of approximately 14 to 18 per cent during the past century. Individual

*Volcanic eruptions are a source of carbon dioxide ($CO_2$) in the atmosphere. Others are respiration by plants and animals, and the combustion of organic materials such as wood, coal, oil and gasoline—the "fossil fuels."*

studies have found increases of from four to more than 20 per cent during this period. It has been theorized that continued increases in the $CO_2$ content could result in a general increase in world-wide temperature within several decades.

Carbon dioxide enters plant tissue through the stomates and in turn appears to have some effect on the size of the stomatal opening. As $CO_2$ increases, stomatal openings tend to decrease in size. If the carbon dioxide around the leaf is significantly reduced, as may occur in calm, still conditions during mid-day, the pore sizes increase. It is not unusual for the $CO_2$ content among growing vegetation, when there is little or no air movement, to drop temporarily to 150 to 200 ppm as a result of the gas being used by the plants. This lowered $CO_2$ concentration may reduce the photosynthetic rate by as much as 10 to 20 per cent below its potential.

Under certain conditions, plants may show a significant response to additional carbon dioxide. Greenhouse experiments have shown mixed results, from virtually no response at all when the $CO_2$ was increased, to yield increases of up to 200 per cent when the $CO_2$ content was doubled. Improved quality and a more rapid maturing of plants has also been noted with increased carbon dioxide concentrations. These variations in results are believed due to the interaction of other factors, including temperature, solar radiation, and moisture.

Generally, the best plant response to increased $CO_2$ has occurred with high light intensities. In other words, if the light is sufficient, then an increase in carbon dioxide will likely produce increased growth. Under normal field conditions, crop canopies are quite likely limited in total yield by both light and carbon dioxide. Whether an increase in the atmospheric $CO_2$ in future years will result in increased crop productivity is quite speculative at this point, since so many other factors are involved. It has been theorized (but in no way proven) that a 30 per cent increase in carbon dioxide would increase the average crop yields by about five per cent, assuming no accompanying change in temperature or solar radiation.

Incidentally, some suggest the observed effects on plants supposedly caused by "talking" to them may be *real* and due mainly to the increased carbon dioxide exhaled by the talker!

Control of carbon dioxide concentrations is important in the storage of fruit and vegetables. As respiration of the stored food progresses, $CO_2$ is released and oxygen consumed. Initially, the increased $CO_2$ and reduced oxygen help to retard the respiration process. However, if the carbon dioxide completely replaces the oxygen, a different type of respiration (anaerobic) will begin, causing the formation of alcohol in the stored fruit. Thus, some type of partially enclosed container which maintains a favorable balance of high $CO_2$ and low oxygen will reduce the rate of respiration and help prevent decay of the stored food. (Tissue or newspapers would act as a "partial enclosure" and help to keep a favorable $CO_2$/oxygen ratio.) Control of the temperature in the storage area is also critical to reduce the respiration rate.

The measurement of carbon dioxide requires expensive and elaborate equipment, and would be of little direct use to most growers. Some greenhouse operations might benefit from additional $CO_2$ in the greenhouse atmosphere. However, anyone contemplating the use of additional carbon dioxide in their greenhouse should make a careful study of the economic feasibility of such an endeavor, and discuss their plans with a qualified Extension Service specialist.

# 4
# Weather and Plant Disease

Weather conditions not only have a direct influence on the growth and development of plants, they also influence plant life indirectly through their effects on plant pests and diseases. Temperature, moisture, radiation, light, wind, air pressure, and evaporation all have an effect on disease organisms and insects. The rate of insect development and the survival and activity of insects depend on weather conditions, as does the rate of disease development and its spread and severity. Plant disease is responsible for several billion dollars in annual crop damage, and the weather is a major factor in much of this loss.

The atmosphere, plants, and disease are closely tied together. The effect of weather on plants (the *hosts* for disease organisms) helps to determine the effect of any disease on those plants. If weather conditions have been favorable and a plant is healthy, a disease will likely have less of an adverse effect than it would on a plant predisposed to disease (or insect injury) by unfavorable weather.

For plant disease to occur, three basic conditions must be present at the same time[1]:

1) A susceptible plant must be in a vulnerable stage.
2) The pathogen (the organism which causes the disease) must be in an infective stage.
3) Environmental conditions must be favorable for the disease to develop and spread.

All three of these necessary conditions are affected, directly or indirectly, by the weather.

# BASIC PLANT DISEASE ORGANISMS

There are four major categories of plant disease organisms—bacteria, fungi, nematodes, and viruses. Common bacterial diseases include fireblight of pear and apple, bacterial blight of beans, black rot of cabbage and other crucifers, and common scab of potato.

Fungus diseases are characterized by the occurrence of spores in their reproductive cycle. The fungi are further subdivided into several classes, but the differences among the various classes are chiefly of interest to plant pathologists and need not concern the ordinary gardener or farmer. Common fungus infections include the "damping-off" disease of young seedlings, various root and stem rots, late blight of tomato and potato, the downey mildews, Fusarium and Verticillium wilt of tomatoes, Dutch Elm disease, powdery mildews of grains and cereals, apple scab, corn smut, and wheat rust.

Nematodes are tiny roundworms found in both soil and water, which cause a number of diseases, the most common of which probably is root knot. This disease affects a wide range of host plants and can cause severe yield reduction or even death. Virus diseases are generally carried by insects, often referred to as "vectors." Some of the more common virus diseases include tobacco and tomato mosaic, cucumber mosaic, bean mosaic, and curly top disease of sugar beets.

| DISEASE OR CAUSE | VEGETABLES |
|---|---|
| *Downy Mildew* | Beans<br>Cucumbers<br>Cantaloupes |
| *Blossom End Rot* | Tomatoes<br>Peppers<br>Watermelons<br>Squash |
| *Root Rot* | Many vegetables |
| *Nematodes* | Cucumbers<br>Cantaloupes<br>Tomatoes<br>Many others |
| *Viruses* | Many vegetables |
| *Anthracnose & Other Leaf Spots* | Beans<br>Cucumbers<br>Cantaloupes<br>Watermelons<br>Peppers<br>Tomatoes |
| *Fusarium & Verticillium Wilt* | Tomatoes<br>Watermelons<br>Cantaloupes |
| *Early Blight or Target Spot* | Tomatoes<br>Potatoes |
| *Late Blight* | Potatoes<br>Tomatoes |
| *Powdery Mildew* | Squash<br>Pumpkins<br>Watermelons<br>Cantaloupes |

| LOOK FOR | HOW TO CONTROL |
|---|---|
| Yellow to dark areas that may be vague or sharply defined, occur on upper surface of older leaves. White to dark cottony mold found on underside of leaves and on bean pods. Vines may be scorched and killed. | Same as for Powdery Mildew, except use maneb on beans and maneb or Bravo (chlorothalonil) on cucumbers and cantaloupes. |
| Blossom end of fruit becomes dry, sunken, leathery and brown-black; or light colored and papery in pepper. One-half of fruit may be affected. | Maintain even soil moisture by watering during dry periods. Mulch or cultivate shallowly during drought. Fertilize properly based on a soil test. |
| Plants may become pale, yellow, and/or stunted. May wilt, die back or collapse. They do not respond to water and fertilizer. Decayed roots may be mushy or firm and covered with mold. | Practice rotation. Avoid wet, poorly drained soils. Remove and destroy affected plants. Control nematodes and prevent injury. |
| Plants may be stunted, yellow, and have galls or swelling on roots. Or roots may be stunted, bushy and discolored. Plants may suddenly wilt in dry weather. | Move garden to a different area or treat soil with a material such as Vapam, Nemagon or Fumazone. |
| Symptoms vary considerably. May consist of yellowing, stunting, spotting, mottling, mosaic pattern, streaking, etc. | Use resistant varieties where possible. Control weeds, insects and nematodes. |
| For anthracnose, check for reddish-brown or black sunken spots or blotches on leaves, stems, pods, and/or fruits. Other leaf spots may vary greatly in shape and color. Leaves may wither and fall early. | Practice two year rotation or longer. Avoid overcrowding and keep weeds removed. Destroy residue after harvest. Spray or dust with maneb or zineb. |
| Seedlings wilt and die. Older plants are stunted while leaves wilt, turn yellow, wither and drop. Dark streaks appear in lower stem and roots. | Plant in well-drained, wilt-free soil and follow a 3-4 year crop rotation. Use resistant varieties where possible. |
| Brown to black spots on leaves and stems. Spots are angular to round with concentric rings (targets). Spots may merge to kill portions of leaves and defoliate the plants, starting at the base. Tomato fruit may sunscald and show sunken leathery spots near the stem. | Practice crop rotation and destroy residues after harvest. Use well-drained soil and keep weeds down. Use resistant varieties. Spray or dust with maneb or Bravo (chlorothalonil). |
| Develops during wet humid season. Irregular, greenish-black, water-soaked spots develop on leaves, petioles and stems. May resemble sunscald or frost damage in advanced stage. A sparse whitish mold appears on underleaf surface. Fruits and tubers also affected. | Same as for Early Blight. See above. |
| White or brown mealy growth on leaves and young stems (especially on the upper surface and occasionally on the fruit). Plants may yellow, wither and die while fruit sunscalds or ripens prematurely. | Use resistant varieties. Practice rotation and good weed control. Space plants well. Destroy crop residue. Spray or dust with Karathane when disease first appears. |

# GENERAL ENVIRONMENTAL EFFECTS

While all diseases are affected by weather conditions to some extent, they do respond differently at various stages in their infection cycle, and each specific disease has its own optimum and lethal conditions. Bacteria require warmth for their initial growth, and free water to germinate. Splashing rainwater is a prime means of disease dispersal—either from one plant to another, or from a plant to the soil and back to one or more other plants. Both temperature and moisture thus are important factors in the severity of bacterial diseases.

Weather conditions are probably even more important to the initial development of fungus diseases than they are for bacterial infections. The fungal spores have exacting temperature and moisture requirements. Dispersal of the spores occurs either in free water or through the air. In some cases, the spores are forcibly ejected into the air and depend on wind currents to spread to new hosts. Once the spores have landed on a new host plant, both temperature and moisture are important factors in the germination and infection cycle.

Since both bacterial and fungus diseases are spread by splashing water, it's advisable to stay out of fields or gardens that are wet from recent rains or irrigation. Movement among wet leaf surfaces will help spread these diseases. Spores may adhere to wet clothing and be rapidly transmitted to other parts of the garden—perhaps to areas that would otherwise be unaffected.

*Wet gardens should be avoided; splashing water is a prime means of disease dispersal.*

Nematodes, while not very sensitive to atmospheric conditions, are affected by soil temperature and moisture. Maximum nematode populations within the soil are generally found when soil temperatures are at or above 70°, while little activity occurs in soils below about 55°. Where soils freeze deeply during the winter, the survival rate of nematodes is quite low. At the other extreme, both the larvae and eggs of this pest are killed by temperatures of about 105° to 110° or higher. Thus, during midsummer, the nematode population in the top inch or two of the soil is significantly reduced.

Viruses are usually not seriously affected by changes in atmospheric moisture since they multiply within the host plant rather than on the plant surface. Air temperature, however, does influence the rate of disease development. Since viruses are usually carried by insects, the effect of the environment on the insects indirectly affects the disease.

In general, then, temperature affects the rate of disease development, while moisture conditions determine whether the germination/infection cycle can be successfully completed. Some pathogens (such as those causing peach leaf curl and powdery mildew of apples) develop best at low temperatures, while others (including those causing Fusarium wilt of tomatoes, various bacterial wilts, and fire blight) require relatively high temperatures for maximum development. Although most disease problems are generally associated with wet field conditions, certain disease organisms are most severe during dry weather. Some of the powdery mildew diseases fall into this category.

## Other Environmental Effects

Winds, both at the surface and up to as high as ten thousand to fifteen thousand feet, play an important role in spore dispersal and the movement of insects. For example, both the cereal leaf beetle and the spruce bud worm may spread over an area following the passage of a weather front, being carried along with the air mass behind the front. South to north movement of the potato leaf hopper often occurs in conjunction with a frontal system. One of the most dramatic examples of spore dispersal is the movement of the wheat stem rust disease northward from northern Mexico and the southern United States each spring. Spores may be carried several hundred miles by the prevailing winds.

Wind, relative humidity, and rate of evaporation also affect plant diseases through their effects on moisture conditions within the crop canopies. Strong winds and low humidity, which lead to relatively high rates of evaporation, greatly decrease the wetting time available to disease organisms on plant or soil surfaces. Dew may fail to form overnight or, if some dew does develop, the plant surfaces will dry rapidly the following morning. Light winds and high relative humidities, on the other hand, will prolong the period of wet foliage well into the late-morning hours.

Even the day length and light intensities can affect the rate of disease development. For example, low light intensities seem to favor development of the powdery mildews and various wilt diseases, while the rust diseases of cereal crops are stimulated by high light intensities.

The type, timing, and method of application of spray and dust fungicides and insecticides are strongly influenced by weather conditions —both past and future. Heavy rains can wash off a pesticide before it has a chance to be effective, while strong winds may prevent a satisfactory application and can lead to significant pollution problems outside the "target" area. Temperatures affect certain chemicals and may significantly reduce their effectiveness. Some insecticides, for example, may become inactive at low temperatures, while others are ineffective in hot and dry weather. Read the labels on these chemical products carefully and adhere to their instructions strictly. Further information on their use and how they might be affected by weather conditions can be obtained from your local county agent or agricultural experiment station.

The use of biological pest controls and organic sprays instead of chemical pesticides removes the risk of inadvertent soil or air pollution, and should be considered seriously as a preferred alternative. However, even these safe methods of disease and insect control can be rendered ineffective by adverse weather conditions, such as heavy rains or extreme temperatures.

# WEATHER EFFECTS ON SOME COMMON DISEASES

More than two thousand years ago, Greek philosophers noted the various disease cycles in crop plants and concluded that variations in the weather caused a spontaneous generation of those diseases.[2] Today we have a much better understanding of the role weather plays in disease cycles, but large gaps in our knowledge remain. One major problem is separating the influence of individual weather factors. Moisture, temperature, light, and other environmental parameters all interact with each other, and it's difficult to determine the exact role of an individual element. Some revealing work has been done in greenhouses and growth chambers, where environmental conditions can be controlled, but it is virtually impossible to match nature exactly. Nevertheless, environmental requirements of certain diseases have been studied sufficiently so that the probability of disease can be forecast in time for growers to take precautionary steps.

The weather/disease relationships that follow are mostly empirical—that is, they were developed from observed data, rather than from some basic physical theory. Therefore, they may not be strictly applicable in all climatic conditions, although they have proven useful in many areas. If you have questions concerning your specific area, or concerning a disease not listed below, contact your local agricultural Extension Service.

## Downy Mildew of Lima Beans

The fungus causing downy mildew is in the same genus as that causing late blight. Again, extended periods of wet, humid weather are required for serious outbreaks of this disease. Conditions favorable to its development are a five-day period with mean temperatures below 79°, minimum temperatures above 45°, and a 10-day rainfall total of 1.20 inches or more. Another forecasting technique, developed at Rutgers University[3], is based strictly on dewpoint temperatures. A period of 40 hours with dewpoints continually above 69° is generally followed by an outbreak of the disease within one or two days. Using forecasts of expected dewpoint temperature, rather than only observed values, alerts can be issued before the end of the 40-hour period, allowing growers to take preventive measures.

## Bacterial Wilt of Sweet Corn

This disease (also known as Stewart's Disease) is carried by the corn flea beetle, whose survival depends on winter temperatures. Mild winters favor its successful overwintering while severely cold seasons will reduce the beetle's population to the point where the disease generally causes no problems the following year. Studies have shown that the *sum* of the mean (average) temperatures of December, January, and February is a good indicator of the severity of this wilt the following growing season, in those areas where the flea beetle normally occurs. If the temperature sum is less than 90°, the disease is usually no problem. A temperature total of 90° to 95° indicates slight wilt development, while a sum of 95° to 100° suggests moderate damage. If the three-month winter temperatures total more than 100°, then a severe outbreak of wilt is likely the following season. Thus, a simple method exists for the home gardener to determine well ahead of planting time the risk of wilt disease injury to his sweet corn crop. If the risk is high, the grower can use a less susceptible variety or plan for a chemical or biological control.

## Late Blight of Potato

Probably the most important disease affecting the potato crop, late blight spreads by spores which develop and disperse under favorable weather conditions. Forecasting procedures for this disease have been developed, including the following:

1) Irish method - blight is likely if air temperatures remain at or above 50° and the relative humidity remains at or above 90 per cent for at least 12 hours, with free water on the plant foliage for at least four hours. If no free water forms on the leaves, then the relative humidity must remain above 90 per cent for at least 16 hours for the disease to begin.

2) Dutch Method - blight will occur within 10 to 14 days if:

   a) dew has developed on the leaves for at least four hours on a given night;
   b) the temperature has not fallen below 50°;
   c) average cloudiness on the following day is at least 8/10 (80 per cent);
   d) at least 0.1 mili-meter (0.004 inch) of rain falls within 24 hours after the period of dew.

3) Beaumont's Rules - blight outbreaks will usually occur seven to twenty-one days after a 48-hour period in which temperatures remain at or above 50° and relative humidities remain at or above 75 per cent.

In general, then, the critical conditions for late blight to develop appear to be a period with air temperatures remaining at or above 50° and high relative humidities (at least 75 to 90 per cent or higher). Free water on the leaves will speed blight development.

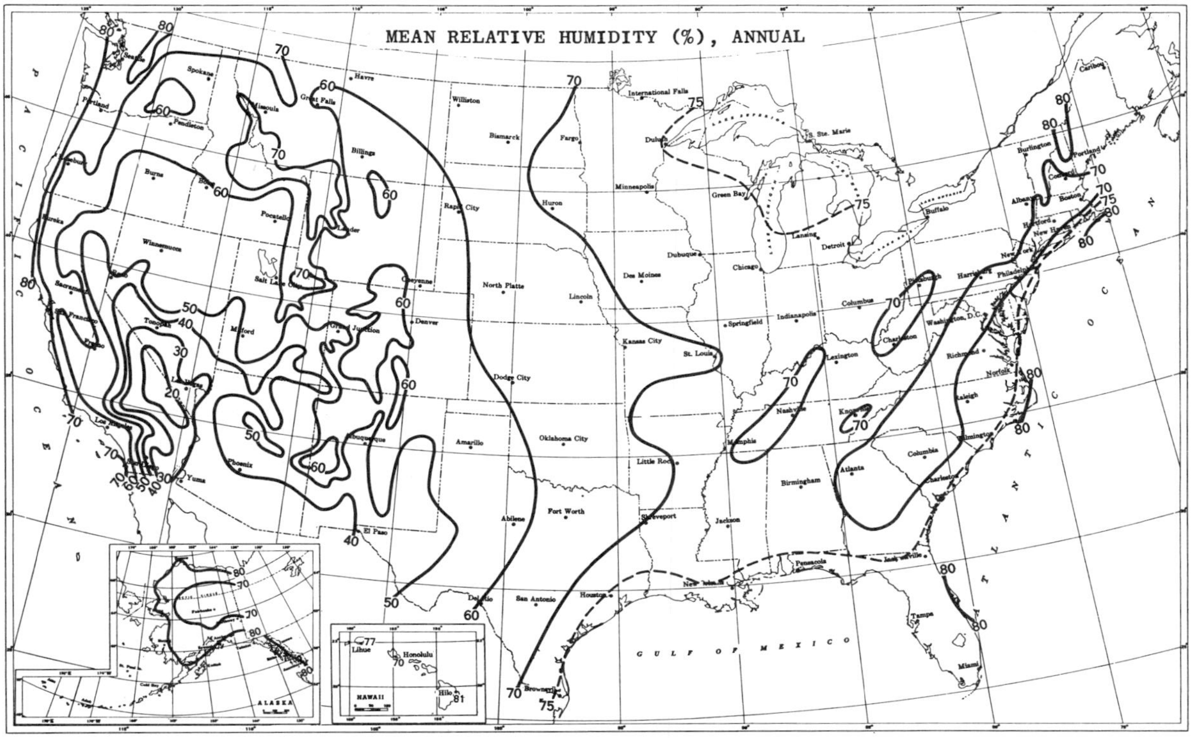


*This chart shows mean relative humidity for the United States, expressed in percentages. High relative humidity is often a precondition for the spread of plant disease.*

## Apple Scab

This is a major apple disease wherever humid conditions frequently prevail. The fungal spores overwinter among dead leaves and other debris on the ground and begin to develop in spring as the weather warms up. The spore sacs swell and burst when wetted, discharging the spores into the air, where wind currents carry them to new hosts. With water on plant surfaces and temperatures of 50° to 75°, germination and infection usually occur within about four to eight hours. The length of the wetting period required for germination depends on the temperature, with the shortest period needed at about 65°. At this optimum temperature, light infection is likely with a wetting period of nine hours, moderate infection with a 12-hour wetting period, and a heavy outbreak of the disease is likely if water on the plant surfaces persists for 18 hours or more.

Apple leaves are most susceptible to scab infection when they are young and become quite resistant as they reach maturity. Thus, warm, dry weather in spring speeds up the development of the host and reduces the severity of the disease. Cool, moist weather has exactly the opposite effect, as it slows down the development of the apple leaves, extending the period of susceptibility.[5]

## Pecan Scab

Damage from this fungus disease is a significant economic factor in the pecan growing industry. As temperatures and moisture conditions become favorable in spring, the spores are spread by wind and rain. The following simple formula has been developed to predict the probability of infection:

$$\text{wetting period (hours)} \times \text{mean temperature (degrees Celcius)} = I$$

For example, a wetting period of 12 hours along with a mean temperature of 50° F. (10° C.) gives an I value of 120 ($12 \times 10$). If I is 140 or above, then infection is probable.[6] This relationship appears to work well until temperatures reach about 80° F. At higher temperatures, and particularly during dry periods, the combination of heat and low humidity becomes increasingly lethal for the spores, and the infection rate decreases rapidly.

## Tobacco Blue Mold

This disease is found mainly in tobacco seedbeds and only rarely in regular tobacco fields in the United States. The time of appearance and the severity of the disease are closely related to January temperatures. Warmer than normal weather during January favors an early appearance and a greater severity of the mold. Once started, the optimum temperature for spore production is about 60°. Periods of cloudy, wet weather with temperatures in the 50s and 60s are extremely favorable for development of the blue mold disease.

## Curly Top of Sugarbeets

This virus disease also affects tomatoes, beans, and cucurbits, along with several other plants. It is found in, and west of, the Intermountain Region of the western United States, but does not extend eastward into the Great Plains. Apparently, the occurrence of this disease is controlled by climatic factors which affect the survival of its vector (the insect carrying the virus), the beet leafhopper.[4]

## Wheat Stem Rust

There's a dramatic northward spread of the spores of this fungus disease each spring. Once the spores have landed, germination can occur within one or two hours at temperatures of about 75° and a relative humidity of over 90 per cent. Full development of the disease is strongly dependent on temperature, requiring about 90 days at 32°, 20 to 22 days at 40°, and only about five days at 75°. Heavy infestations of this disease are most likely when average temperatures remain at or above 60° and the relative humidity remains more than 90 per cent for extended periods.

## Southern Corn Leaf Blight

This fungal infection was considered a relatively minor corn disease until 1970, when a new strain developed and caused severe damage throughout the corn-growing region of the United States. The disease is favored by rainy weather, persistent dews, generally high relative humidities which maintain moisture on the leaf surfaces, and temperatures ranging from about 55° to 80°. Spores may be carried by winds for long distances, while local spreading from plant to plant occurs with rain splash. A film of water on a plant surface for six to eight hours is sufficient for germination and infection to occur. The entire infection cycle may be completed within 60 hours if conditions are favorable.[7]

## Other Diseases

Cotton anthracnose, which produces lesions on affected plant parts, especially leaves, is a significant disease in the more humid cotton-growing regions of the southern United States, but virtually absent in drier areas. A line representing an average summer rainfall of approximately 10 inches separates the disease-free regions (those on the dry side of the line) from those areas in which the disease is prevalent (on the wetter side).[8]

Onion smut appears to be limited to those areas where soil temperatures remain below about 77° F. during the first month or so of seedling growth. The optimum soil temperature for the disease organism is 55° to 72°. As temperatures increase above this optimum, the percentage of disease germination drops sharply. Thus, by growing onions in late summer or early fall when soil temperatures are near their highest, the problem of smut can be reduced significantly.

The rootknot nematode* is sensitive to soil temperatures, soil moisture, and even light. Optimum soil temperature is about 68° to 85°, with activity virtually ceasing below about 55°. Optimum soil moisture is between 40 and 80 per cent. Most nematodes reach a maximum population when soil temperatures are about 70° or higher.

Tomato transplants appear to be less susceptible to wilt diseases if they are initially grown at high soil moisture contents and at temperatures either higher or lower than those usually considered to be optimum.

Both angular leaf spot of cotton and black rot of cabbage and other crucifers are most severe on plants that have been predisposed by cool, damp spring weather. Delaying planting until later in the season when more favorable conditions occur will result in more rapid early growth of the plants and better control of the disease.[9] Here, then, is one important reason to resist the urge to plant too early in the spring, especially in middle and northern latitudes. In addition to the hazard of a late freeze, most plants are predisposed to disease injury by cool, excessively wet soils—conditions which often prevail during early and mid-spring. Slow-growing, weak plants are much more susceptible to disease organisms than are healthy, vigorously growing plants.

Other disease/weather relationships may be found in the references listed at the end of this chapter and in appropriate references found in Appendix D.

*Root-knot nematodes are a species of nematode which produces the disease known as root knot—a disease characterized by galls which form on the roots and tubers of infected plants.

# 5
# Frost and Other Weather Hazards

There are several atmospheric "events" that play important roles at one time or another in crop production. These events may be classified as weather hazards—the principal ones being frost and freeze damage, hail, wind, and drought.

Drought and wind may threaten agriculture in virtually any region of the world. Much of the world-wide food crisis of the early 1970s resulted in large part from drought in parts of the southern Sahara and the failure of monsoon rains in India and other Asian nations. Severe drought has stricken California and other parts of the western United States although heavy rains during the 1977-78 season ended the problem, at least temporarily.

Frost and freeze damage, although absent in tropical regions and rare in most of the subtropics, affects sizable agricultural production areas around the globe. Hail is most likely to occur in certain "preferred" regions (especially central North America and parts of Russia), but it devastates other areas as well from time to time.

# FROST AND FREEZING INJURY

One of the most widespread and persistent hazards to crop growth and production in mid- and northern latitudes is frost. Frost probably causes more damage to home gardens (and more frustration to gardeners) than any other weather hazard. In southern areas, on the fringe of the "frost belt," dangerously low temperatures are most likely to occur during the winter growing period, while in other sections the critical periods are spring and fall, near the beginning and end of the growing season.

Spring frosts threaten newly emerged seedlings and developing fruit tree buds, while fall frosts can bring an early end to the natural growth cycle of plants. Severe later-than-usual spring frosts can destroy a developing crop, often necessitating replanting. Premature fall freezes, while also causing serious injury or death, generally result in reduced yield, but seldom a complete loss.

Confusion often arises over the terms "frost" and "freeze." The white covering of ice crystals which is observed on plant and other surfaces is known as frost, and develops similarly to dew, except that the plant surface temperature is 32° F. or lower and ice crystals develop rather than water droplets. When the plant's surface temperature cools to the dewpoint of the air, moisture from the atmosphere condenses on the plant surface. The plant's temperature at the time of this condensation determines whether water droplets (dew) or ice crystals (frost) develop.

Because sharp differences in air temperature can exist near the ground, and also since plant temperatures may be several degrees below the temperature of the surrounding air, frost frequently occurs even when temperatures inside a weather instrument shelter remain above 32°. The difference between a plant and shelter temperature depends on various factors such as wind speed, cloud cover, plant height, and local terrain features. Frost (sometimes called white frost, hoar frost, or radiation frost) generally occurs with clear skies and very light or calm winds—ideal conditions for the loss of heat by longwave radiation from the earth's surface (hence the name radiation frost).

Under certain conditions, plants may be damaged or killed by freezing temperatures without frost being observed on the plant surface. If the air is quite dry (and therefore has a very low dewpoint) the plant surfaces may not cool to the dewpoint and no condensation will occur. However, if the plant temperature falls below the freezing level, severe damage can result even though no frost crystals are formed on the leaf surfaces. This phenomenon is sometimes referred to as "black frost," because affected vegetation often blackens and shrivels.

Temperatures of 32° or lower can also occur with no observed frost on plants during "advection" freezes. These freezes are caused by a cold mass of air moving into a region, and often occur with strong winds and considerable cloud cover.

In general, then, *frost* refers to a noticeable white covering of ice crystals on plants (or other exposed objects), with shelter temperatures either above or below freezing, while a true *freeze* indicates an actual air temperature of 32° or lower, with or without frost. Until 1948, the U.S. Weather Bureau (National Weather Service) recorded the first date in fall and last date in spring of a "killing frost." Since then, however, official weather records list the earliest and latest dates of specified shelter temperatures, usually 32°, 28°, 24°, 20°, and 16°. The lower temperatures are critical levels for various crop and orchard plants at different stages of development.

*This heavy ice covering resulted from an unusual storm in Boulder, Colorado. The photograph was taken by the National Center For Atmospheric Research in Boulder.*

## How Freezing Kills

Whether actual frost is observed or not, freeze damage can cause serious injury or death to plants. Such injury can occur in several ways. Normally, freezing temperatures cause the formation of tiny ice crystals within the spaces between individual cells of plant tissue. As these ice crystals grow, they "feed" on water which moves from within the cells, across the cell membrane, to the crystals. This loss of water within the cell causes a dehydration of the protoplasm and, if severe enough, death of the cell. At the same time, the sharp edges of the ice crystals cause mechanical injury to the cell protoplasm.

If freezing is quite rapid, ice crystals may form *within* the cells as well as in the intercellular spaces. These small crystals rip and tear the cell material, eventually killing the cell. Thus, individual cells and the plant itself are injured or killed by a combination of mechanical damage due to the ice crystals and a loss of water within the cells. Additional injury may result if rapid thawing occurs. In this case, water reenters the cell too quickly and the protoplasm is distorted and damaged.

A different type of mechanical injury occurs to plants when periods of freezing and thawing lead to soil heaving. In this case, the movement of the soil causes roots to be damaged or, in some cases, torn away from the plant. Not only does the damaged root system retard growth and development of the plant the following growing season, but it may also lead to winter killing of the plant.

Winter-kill occurs when atmospheric conditions favorable to rather high transpiration rates exist at the same time that water absorption through the roots is limited—either by a damaged root system or by frozen or extremely dry soil. One of the prime reasons for applying a winter mulch to trees, bushes and other perennial plants is to reduce the depth of soil freezing and to prevent the heaving of the soil. It is also a good idea to make sure that the soil is moist before freezing begins in the late fall or early winter.

Some plants, especially tropical or semitropical species, may show chill injury at temperatures above freezing. This type of injury is caused by physiological changes within the plant tissue rather than by ice crystal damage. Often, such injury may not be permanent, but retards growth and development of the plant and hence reduces yield. Tomatoes, for example, will not ripen properly if air temperatures drop below 40°. True tropical fruits, such as bananas, coconuts, mangoes, papayas, and coffee, may show injury if temperatures dip much below 50° F.

## Plant Hardiness

As almost any gardener realizes, not all plants are affected in the same way by a frost—unless, of course, there is a severe freeze with temperatures dropping well below freezing, after which virtually all growth ceases on perennial plants and all annual plants are killed. The ability to withstand freezing temperatures is referred to as frost (or cold) hardiness. Hardiness not only varies among species, but also varies for a given species depending on the time of year and pre-conditioning. In general, as the temperature gradually decreases in the fall, hardiness increases in perennial plants. With fruit trees, for example, hardiness increases during the fall and winter months, then begins to decrease in early spring.

Seedlings for spring planting in the vegetable or flower garden may usually be hardened by gradual exposure to temperatures just above freezing. Cold frames are often used for this hardening-off process.

The exact way in which this hardening-off permits plants to tolerate low temperatures is not fully known, but apparently it involves structural changes in the cell protoplasm which enables plants to better withstand mechanical stresses. Also, those plants which are most tolerant of cold appear to have a greater percentage of their cellular water chemically or physically bound to the cell and therefore not subject to crystalization. In addition, their cells generally have a lower water content and a higher concentration of sugars, which lowers the freezing point of the water—acting as a plant's "antifreeze."

*A good way to "harden-off" vegetable plants or prepare them for outdoor conditions is to use a cold frame.*

## Critical Temperatures

It is extremely difficult to pinpoint the exact temperature at which damage occurs within a given crop. It depends upon such variables as stage of development, general plant health, and soil conditions. Also, the length of time the low temperatures persist and the rate of fall of the temperature are important. Injury or death more often occurs at higher temperatures during rapid cooling than when the cooling is more gradual. In addition, in recent years there has been a profusion of new varieties of many plant species, each with a unique cold tolerance.

Tables 8 and 9 are a general guide to critical temperatures for a variety of vegetables and fruits. No attempt has been made to account for varietal differences, but the values can be used as guidance in determining if particular crops should be protected. Growers should check with seed suppliers to see if specific temperature tolerances have been determined for varieties they are growing.

The data in the tables are compiled from various sources and thus cover a range of varieties and growing conditions. In Table 9, the temperatures listed are those that are critical early in the growing season as buds begin to open and the fruit develops. For most fruit trees grown in temperate regions, the dormant trees themselves can stand winter temperatures of 10° to 20° below zero, or even colder, depending on the variety and degree of maturity.

## Table 8: Critical Low Temperatures (°F.) for Initial Frost Damage to Vegetables*

| Crop | Temp. | Crop | Temp. | Temp. | Crop |
|---|---|---|---|---|---|
| Artichoke | 31-32 | Garlic (dry bulbs) | 29-30 | Spinach | 30-32 |
| Asparagus | 30-31 | Horseradish (roots) | 29-30 | Squash (summer) | 31-33 |
| Beans (pods) | 31-32 | Kale | 27-29 | Squash (winter) | 30-32 |
| Beets (roots) | 29-30 | Muskmelon | 33-34 | Sweet corn | 32-33 |
| Beets (tops) | 31-32 | Okra | 29-30 | Sweet potatoes | 32-33 |
| Broccoli | 29-30 | Peas (pods) | 28-30 | Tomatoes (fruit) | 32-34 |
| Cabbage | 26-28 | Potato tubers | 28-30 | Watermelon | 32-33 |
| Carrots | 28-30 | Pumpkin | 31-32 | | |
| Cauliflower | 27-29 | Radish | 30-32 | | |
| Celery (stalk) | 31-32 | Rhubarb | 30-32 | | |
| Cucumber (fruit) | 30-32 | | | | |

*At these temperatures at least some effect is noted on the quality and/or quantity of the yield. More tender plants may be killed, while others will survive but produce poorer yields.

## Table 9: Critical Low Temperatures (°F.) for Frost Damage to Fruits

| Fruit | Buds Closed No Color | Buds Showing Color | Full Bloom | Small Immature Fruits |
|---|---|---|---|---|
| Almonds | 20-24 | 23-26 | 25-28 | 29 |
| Apples | 23-26 | 24-26 | 27-30 | 29-30 |
| Apricots | | 25 | 26-28 | 30-31 |
| Blueberries | 25 | 25-26 | 26 | 27-28 |
| Cherries | | 28 | 28-30 | 30 |
| Grapefruit | | 26-28 | 28 | 28-30 |
| Grapes | | 30-31 | 31 | 31 |
| Lemon | | 26-28 | 28 | 29-30 |
| Limes | | 26-28 | 28 | 30 |
| Orange | | 26-28 | 28 | 29-30 |
| Peaches | | 25 | 26-28 | 29-30 |
| Pears | 23-25 | 26-27 | 28-30 | 30 |
| Plums | | 24-26 | 26-28 | 29-30 |
| Prunes | | 23-26 | 27-29 | 29-30 |
| Raspberries (red) | 25 | 26-28 | 29 | 27 |
| Raspberries (black) | 26 | 27-29 | 30 | 28 |
| Strawberries | 22 | 24-28 | 31-32 | 28 |
| Walnuts (English) | | 30 | 30 | 30 |

# FROST PROTECTION FOR HOME GARDENS

Over the years, many different techniques have been developed for protecting crops from frost and freeze injury. Methods often used by the home gardener or small-scale farmer may not be suited to the commercial grower, and vice versa. Equipment such as heaters, smoke generators, sprinklers, large wind machines, and even helicopters are employed by commercial growers, but these techniques are too expensive or impractical for the back-yard gardener. On the other hand, methods such as covering individual plants with baskets or hotcaps, or making cloth "tents," while suitable for small plots, are not feasible for the grower with many acres of vegetables.

Thus, each grower must determine his own individual needs, balancing costs and time involved against the value of the crop. There are numerous ways in which home gardeners and small-scale farmers can reduce, if not eliminate, the risk of frost damage.

## Cultural Practices

One of the simplest ways to reduce frost risk in spring is to hold off planting operations until the probability of frost is low (see Chapter 8). During the first few mild, sunny days in early spring, virtually every red-blooded American gardener is struck with a tremendous urge to go forth and sow. In some areas there is strong competition to have the first peas, or the first ripe tomatoes in the neighborhood. If a late frost hits, however, the "slow" gardener may not only have the first, but the *only* peas, corn, tomatoes, etc. At best, early-bird gardeners may have to replant parts of their gardens, causing their crops to be later than if they had held off planting a while longer. For commercial growers, the risk of a late frost may be outweighed by the potential for higher prices normally obtained for the season's first fresh vegetables. But the average home gardener has nothing to lose by being a little cautious.

Information on frost probability dates is available in most states through the agricultural experiment stations, land-grant colleges, or the county Extension Service. These publications often contain additional information on frost damage and protection, and are useful guides.

While the dates of various frost risks in the spring provide a guide to planting times, similar information for fall frosts can enable you to plan your harvest dates to avoid undue risk of freeze damage before the crop has matured. The spring and fall frost-risk dates together determine the length of the growing (or frost-free) season. With this information, and a knowledge of the calendar-day or growing-degree-day (see Chapter 6) requirements of a particular crop, you can select varieties likely to mature within the normally available frost-free period in your area.

The selection of a garden or orchard site is very critical. Since cold air drains downhill, low-lying areas tend to be the coldest during radiation frosts. Thus, planting sites on higher areas, or on the upper portions of slopes, will generally be safer from frost than those on the lower parts of slopes or in valley bottoms. Even shallow depressions on relatively flat land may cool several degrees lower than surrounding high spots.

Frost is less likely to occur over bare soil than over a sod cover, and tilled land has a greater frost risk than untilled soil, under similar weather conditions. The reason in both cases is the same. During the day, heat penetrates untilled soil deeper than plowed land, and bare soil deeper than sod-covered soil. At night, this stored heat is conducted upwards toward the surface and helps to warm the air next to the soil. Tilling breaks up the soil and produces many "air pockets" which are very poor conductors of heat. With sod, air is trapped near the ground, again resulting in poor conduction of heat. Thus, less heat is stored in the ground during the day, and the heat that is available at deeper levels is not conducted upward very readily at night.

Likewise, a mulched soil presents a greater frost risk than an open, unmulched plot. Thus, a clean, open, untilled soil surface will help minimize frost damage. Delaying the application of mulch in the spring until danger of frost is past will not only help to reduce the risk of frost damage, but the resulting warmer soil will enhance plant growth. The one exception to this rule is the use of black plastic as a mulch. In this case, the black plastic lies flat on the soil surface, producing few air pockets to retard heat conduction. In addition, the black color of the plastic absorbs a maximum amount of the sun's heat and transmits it to the soil. This type of mulch is therefore used rather early in spring to promote a more rapid warming of the soil, permitting an earlier planting of warm-season crops.

Of course, if the temperature drops well below freezing, then damage will be extensive regardless of surface conditions. However, for those times when temperatures are just around the critical level (and this is usually the case for the last one or two spring frosts and the first one or two frosts in the fall), proper cultural techniques and site selections can make the difference between losing or saving a crop. In many sections of the country, the first critical low temperatures in the fall are often followed by several weeks of fine "Indian Summer" weather. Thus, by protecting your plants enough to survive this first threat, your growing season may be extended significantly.

*A gardening site with a gentle slope allows cold air to drain down hill, away from plants. A tree line provides wind protection.*

## Mechanical Protection

Many different techniques are used by home gardeners to protect tender plants early in the spring and again in the fall. Almost any type of cover will help to hold in the heat from the ground and maintain higher temperatures than those outside the cover. I have been quite successful putting old sheets or blankets over stakes to cover threatened plants. Others use baskets, hot caps and cloches to protect individual plants or plant groupings. Temperatures inside such coverings will remain about two to six degrees above outside air temperatures, depending on local conditions. That extra warmth is often sufficient to protect threatened plants during critical temperature periods.

These coverings should be removed during the day. Even glass or plastic covers which allow sunlight to penetrate can be detrimental to the plants. The combination of increased daytime heat and humidity within the covers can be harmful. Also, the higher daytime temperatures may reduce the frost hardiness of the plants.

You should keep in mind that virtually all of the frost protection methods described above are useful mainly for radiation frosts—those frosts that occur with clear, starlit skies and calm or very light winds. Advection freezes, often accompanied by rather strong winds, are extremely difficult to protect against. Some protection may be afforded by walls or other windbreaks and sturdy covers, but, in general, a severe advection frost (and a really severe radiation frost, too, for that matter) will take its toll unless some form of artificial heat is avail-

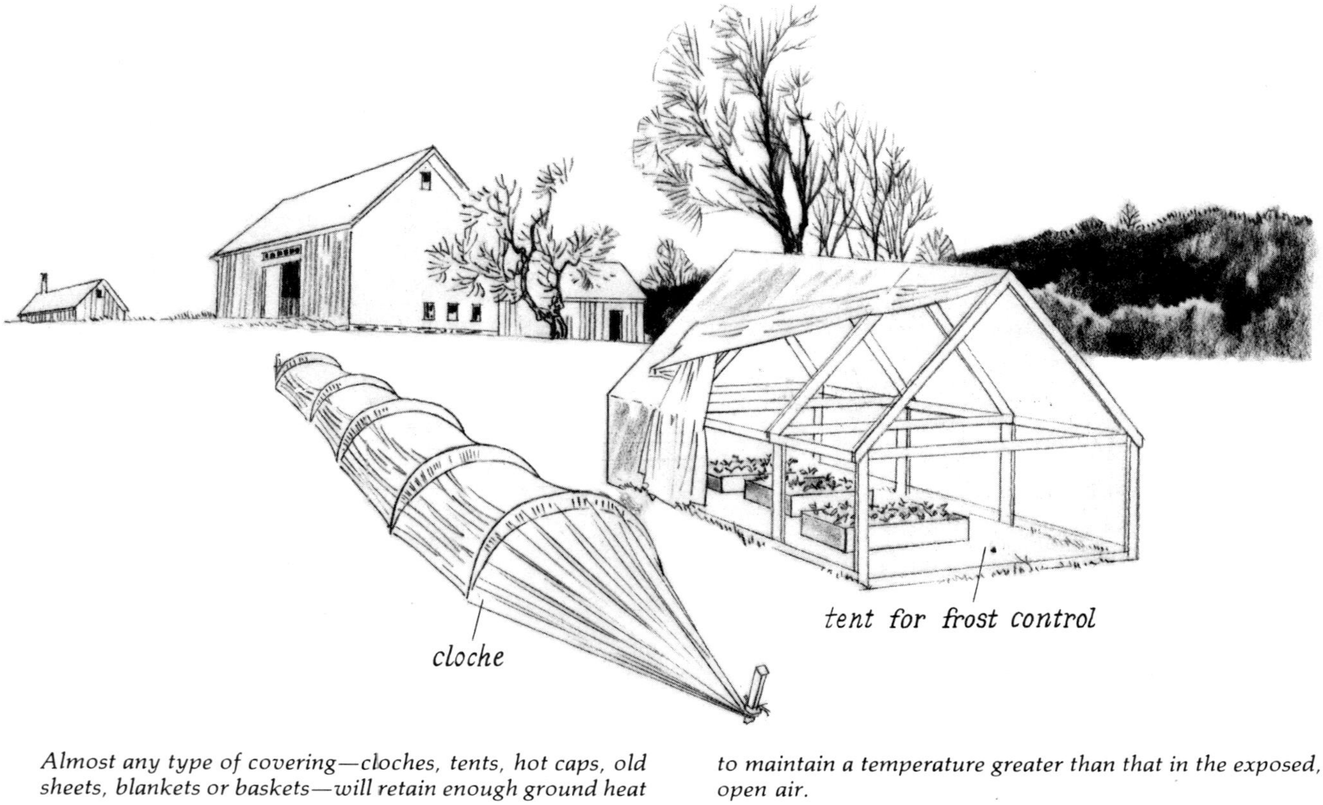


*Almost any type of covering—cloches, tents, hot caps, old sheets, blankets or baskets—will retain enough ground heat to maintain a temperature greater than that in the exposed, open air.*

able. Usually, however, the final spring and the first fall frosts are the radiation type which can be protected against.

# FROST PROTECTION (LARGE SCALE)

## Heaters

A variety of frost protection measures are available for large scale farming or agricultural operations. For example, heaters have been used over the years, and the majority in use today are oil-burning models. In some areas, anti-pollution regulations may now prevent or severely restrict their use. The basic purpose of a heater is quite simple—to promote sufficient heat to maintain air and plant temperatures above critical levels. Numerous small heaters are usually preferable to a few large ones; typically, about 25 to 35 heaters per acre are used. Heat is transferred both by convection (upward movement of warm air) and radiation (radiant heat). The latter becomes the major form of protection when windy conditions prevail. In hilly terrain, heaters are generally placed in low spots, allowing the heat to rise along slopes.

## Wind Machines

Large wind machines, or fans, are appropriate during those nights when conditions are most favorable for radiation frosts, and a rather strong temperature gradient exists with the coldest air near the ground and warmer air aloft. This condition is known as an inversion.

Under inversion conditions, since cold air is more dense (heavier) than warm air, the air becomes "layered." Warmer air above an orchard, for example, will tend to remain where it is and not mix with colder and denser air lying at the surface. The purpose of the large fans is to force the warm and cold air to mix, thus raising the temperature in the lowest layers near the ground, where crops are located.

Wind machines for frost protection were first used about 1920. Although the equipment has changed since then, the basic technique remains essentially the same. Usually one or more fans are placed at roughly 30 to 40 feet above the surface. Although the technique varies with topography and crop type, about one machine per every ten acres to be protected gives satisfactory results in most cases. The machines are effective only if the air at the fan height is about 10° to 15° warmer than the air at the surface on those nights when frost threatens. At best, the use of fans may increase surface temperatures by about six to eight degrees.

Since the wind machines are expensive, it is good practice for those considering their use to first make temperature measurements for at least one full season to determine if the necessary conditions are commonly found in the particular place of interest. Various types of thermometers can be placed at different heights on poles to record the temperature conditions at and above the surface. If a 10° to 15° temperature difference between the shelter and the projected fan height does not occur frequently on nights with good radiational cooling, then wind machines are not feasible. Also, fans will not be useful during advection freezes when winds are brisk and the vertical temperature gradient is small.

In some cases, a combination of fans and heaters provides the best protection from frost. Usually, fans are used initially until they can no longer provide adequate warming, and then heaters are turned on.

An expensive but effective variation of the large fan technique is the use of helicopters. One or more of these aircraft are employed to mix the air over threatened cropland. Each helicopter can protect about 10 to 12 acres, depending on temperature and wind conditions. Commercial firms around the country provide this specialized service, and can be located through your county agent or through advertisements in agricultural publications.

*Thirty-foot high wind machines pull higher warm air down into orange trees during cold winter nights in California's San Joaquin Valley.*

## Sprinklers

Another method of frost protection employs the use of sprinklers to apply a constant spray of water on threatened crops. This technique is appropriate because heat is released when water changes from a liquid to a solid state (ice). This release of heat will keep the surface temperature of a plant near 32° as long as water is applied continuously.

If done correctly, the technique can be quite satisfactory. There are, however, several pitfalls. In the first place, it is absolutely necessary to maintain a layer of liquid water on the plant or ice surface. Otherwise, with no water to freeze, the release of heat is stopped and surface temperatures fall below freezing rapidly. Another danger: Excessive ice formation which leads to limb breakage and other plant damage. Rates of about 0.05 to 0.30 inches of water per hour are generally used, depending on weather conditions. With the higher rate, the risk of plant damage due to heavy ice loads is increased. Because of the high rates of sprinkling required to protect against extremely low temperatures, this technique is best suited to those times when air temperatures are about 23° to 25° or higher.

Sprinkling should start when the air temperature approaches the critical value and must be continued until either all the ice has melted, the ice has begun to melt and water is dripping freely from the surface, or the wet-bulb temperature has gone above 32°. No matter how successful a sprinkling operation has been over night, the entire effort can be wasted if sprink-

*This is a frost control system used in Florida. Some of the sprinklers froze and would not rotate, causing excessively heavy ice build ups.*

ling is stopped too soon the following morning. This point cannot be overemphasized! The greatest danger is stopping the sprinkling prematurely.

One other potential danger with sprinkling: The cooling that occurs when water evaporates. This cooling effect of evaporating water is usually not important with calm or light winds. However, as the wind increases, so do evaporation rates, and sprinkling is therefore not recommended when wind speeds exceed about four to six miles per hour.

## Other Large Scale Protection Methods

Other commercial frost protection techniques, in addition to the use of heaters, wind machines, helicopters, and sprinklers, have been used, or at least tried.

Flooding as a means of frost protection is common in cranberry bogs and has limited use elsewhere. The water retains sufficient heat to maintain plant temperatures above critical levels. Caution must be exercised to avoid excessive or too frequent flooding which can drown root systems and cause other soil/plant problems. Used properly, this method can be very effective.

The use of "smoke screens," created by burning various substances to produce smudge fires, has generally fallen into disfavor—mainly because air pollution problems may result. Not only do new clean-air standards prohibit the use of smudge fires in many areas, but research indicates that, in some cases, the resulting pollution may have harmed the vegetation that the fire was supposed to protect.

Another addition to the array of "frost fighters" has been microwave radiation. The use of microwaves in frost prevention is quite new and requires additional testing, but does hold some promise of a clean, but expensive, technique. Plants absorb the microwave radiation which produces warming within plant tissue. Radiation can easily be directed to specific plant parts—for example, blossoms or developing fruit.

# FORECASTING FROST

As most growers know, the first frosts in fall and last ones in spring are often spotty—some fields and gardens are hard hit, while others nearby seem to be unaffected. Even various parts of the same plot may show different effects. Of course, if there is a "hard" freeze and temperatures dip into the mid-20s or lower, or a strong advection frost has occurred, then all vegetation will be affected. But when temperatures just fall to the critical level, say the low to mid-30s, then considerable variation in frost damage can occur because of local differences in the micro-climate (see Chapter 9).

Under such critical conditions, will frost occur in your particular plot? If growers had a good method of determining whether frost was, or was not, a distinct possibility at their location, much needless labor could be avoided and many crops saved. While there is no ideal method for frost prediction, several techniques can be useful for estimating the likelihood of local frost. Again, we are concerned with radiation frosts, since advection freezes usually cover large areas with very little local variation, and are predicted in general weather forecasts.

Remember that radiation frosts occur with clear skies and very light or calm winds. Often these conditions prevail the first or second night after a cold front passes, as a high pressure area settles over the region. In this situation, temperatures usually begin to fall steadily around sunset and continue to drop, more or less steadily, throughout the night. The most rapid drop usually occurs during the first several hours after sundown.

General weather forecasts are usually issued for large areas and cannot pinpoint exact locations of frost. In fact, it is virtually impossible to do so, unless temperatures will be so low that all areas are likely to experience a freeze. About the best that can be done is to issue a frost or freeze warning, indicating the low temperatures expected overnight and whether the freeze will be general or confined mainly to "normally colder areas."* It is then up to individuals concerned about frost to be familiar enough with their specific locations to determine whether frost is likely or not, given the expected "official" low temperatures.

By making regular observations of the air temperature at your location over a period of time, you can develop your own method of "forecasting" radiation frosts. When frost is a possibility, record the temperature at some time in the evening (say 7 or 9 P.M.) and then note whether or not frost occurs the next morning.† As you acquire a series of data you will begin to notice that temperatures below a certain value at the evening observation time are usually followed by frost the following morning, while temperatures above a particular value seldom, if ever, are followed by frost. In between these upper and lower values will be a range of temperatures which may or may not indicate that frost is likely. I have found, for example, that in south-central Pennsylvania if evening temperatures around 7 P.M. are in the low-40s or lower, with clear, calm conditions expected overnight, then some frost is likely by morning. If the 7 P.M. temperature is in the upper-40s or higher, frost is not likely. In the middle 40s—take your pick!

This method isn't foolproof, but it is simple and you'll be right more often than not. While actual temperature values will vary somewhat from one region to another, the general

*In a few areas of the country, special agricultural or fruit-frost forecasts are issued. See further discussion in Chapter 7.

†To be completely accurate, you should adjust for any change between standard and daylight saving time. Thus, if you take your 7 P.M. temperature as a guide to frost risk during standard time, you should use the 8 P.M. daylight-time temperature as your guide. In this way you use the same "sun" time.

method is applicable anywhere. It is a good idea to develop separate temperature criteria for the critical spring and fall seasons. Differences in daylength, soil temperatures, and general weather patterns can produce differences in the evening-temperature/morning-frost relationship between the two seasons.

Another, more involved (but also more reliable) variation of this technique is to relate frost to the change in temperature during a certain time period the previous evening. For each night when clear, calm weather prevails and there is a threat of frost, plot a point showing the actual temperature at 7 P.M. and the change in temperature between 7 and 10 P.M. The following morning note whether a frost occurred and indicate next to the plotted point either a F (frost) or a N (no frost). Collect data for at least one fall and one spring season during the period when frost is of concern in your area. Plotting a separate graph for spring and one for fall will give better results than one graph using data for both seasons combined. Then look closely at your graphs. You will probably be able to pick out an area with all "Fs" and an area with all "Ns." In between will be a region with both "Fs" and "Ns." Draw lines to separate these areas and you will have yourself a reasonably good guide to forecasting frost at your location. The next time frost is a threat, note the 7 P.M. temperature and the change in temperature between 7 and 10 P.M., enter the point on the graph and see which section it lies in. If it is in the questionable section you will have to decide whether the potential loss is worth the effort of protecting your plants. Use these additional points that you accumulate with time to update your graph and improve its reliability. Five years of data, for example, will produce a more reliable graph than one based on only one or two years of record.

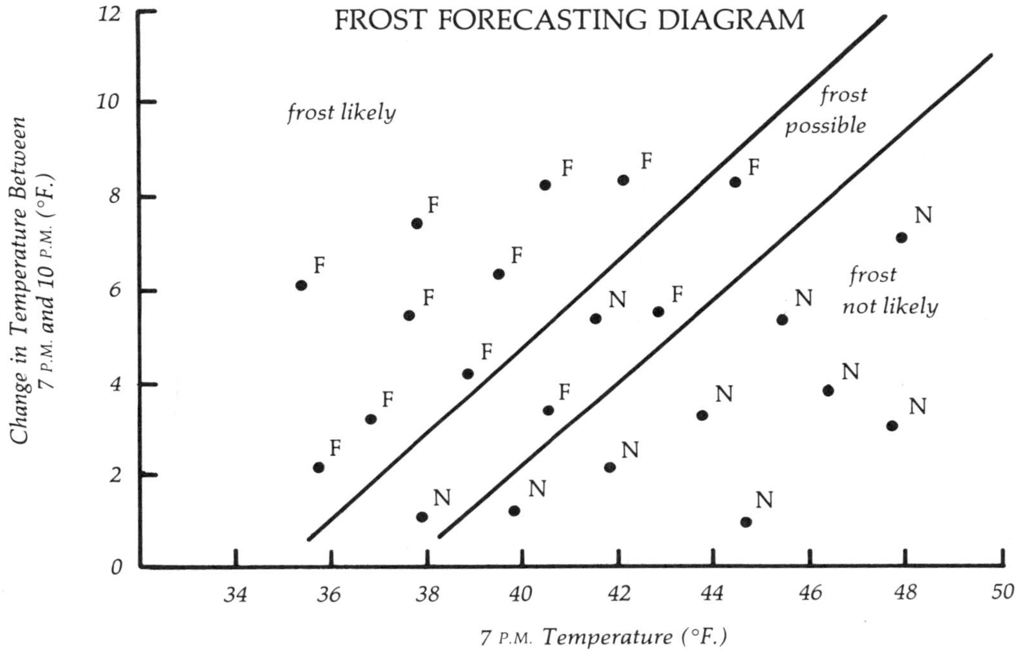


This example is given simply to illustrate a particular technique. There are no hard and fast rules as to its use, and you may discover a somewhat different variation that works best for you.

and 10 P.M., enter the point on the graph and see which section it lies in. If it is in the questionable section you will have to decide whether the potential loss is worth the effort of protecting your plants. Use these additional points that you accumulate with time to update your graph and improve its reliability. Five years of data, for example, will produce a more reliable graph than one based on only one or two year's of record.

This example is given simply to illustrate a particular technique. There are no hard and fast rules as to its use, and you may discover a somewhat different variation that works best for you.

Many different techniques for predicting frost have been developed by agricultural research groups, grower's associations, and other organizations. Generally, their methods are valid for their own particular location, but they may be adapted for other areas. Many of these techniques use dewpoint temperature in addition to the regular air temperature, since the amount of moisture in the air is one factor in determining the rate at which the temperature falls overnight—the drier the air, the greater the temperature drop, other things being equal. The use of these methods involving dewpoint temperatures is therefore limited to those persons having psychrometers or hydrothermographs (see Chapter 1), but since relatively inexpensive instruments can be purchased to determine the dewpoint, two of these techniques will be briefly described.

Probably the simplest technique uses the 9 P.M. dewpoint temperature. If the dewpoint is 36° F. or lower, frost is considered likely and warnings are issued. Again, the actual value may be slightly different at your location, but you can soon determine this for yourself. In any case, it is not likely that the critical dewpoint for your area would vary more than a degree or two from the 36° value cited above. The "Bliss Formula," developed for use in the New Jersey cranberry bogs, uses the air temperature ($T_a$) and dewpoint temperature ($T_{dp}$) one-half hour after sunset. The sum of these two temperatures multiplied by 0.31 produces a forecast of the lowest minimum temperature likely in the bogs overnight. The proper formula reads:

$$0.31(T_a + T_{dp}) = \text{minimum bog temperature (°F.)}$$

For growers who cannot measure the dewpoint temperature, a simple method of frost prediction is to note the air temperature when dew first begins to form on the grass or other vegetation in the evening. If the temperature is above 40°, the danger of frost is slight. If the temperature is below 40° there is a good chance of some frost, and the risk becomes greater the lower the temperature at which the dew first forms.

# DROUGHT

While most weather hazards such as frost, hail, strong winds, and flash floods, cause virtually instantaneous damage to crops, drought is a long-term, initially almost imperceptible, danger. At first, the sunny, warm days are welcome, allowing much time for doing the necessary outside chores and providing the warmth and radiant energy needed for plants to grow and develop. But gradually, after a prolonged period of dry and perhaps hot weather, the lack of moisture begins to tell on the crops. Exactly how long a dry period is required to cause detrimental effects on crop growth will depend on the amount of moisture initially in the soil, the type of soil, crop type and density, and general cultural practices.

*When most severe, drought can produce dust storms and scenes like the one shown in this classic photograph from the 1930s.*

At the beginning, the dryness may be noticeable only for a few hours during the mid-afternoon, when temperatures are at their highest, and solar radiation is near its peak. The plants are unable to absorb enough moisture from the soil to balance the water lost through evapotranspiration, and incipient wilting begins. Leaves may begin to curl and the plant loses its vigorous, healthy look. The effects of drought have begun.

Overnight, cooler temperatures and perhaps some dew permit the plants to recover and by morning they seem back to normal. Each day the period of wilting becomes longer, however, and, unless additional moisture is provided, serious damage or death will result. Even when the plants recover from a serious drought, both the quality and quantity of yield are usually reduced.

Drought is no stranger to any agricultural region, although some areas are more susceptible than others. Lands bordering permanently arid or semi-arid areas run the greatest risk of severe drought losses to crops. The worldwide food crisis in the early 1970s was primarily due to the failure of seasonal rains in many of the marginal agricultural regions of Asia and Africa. Even normally humid regions occasionally experience prolonged and damaging dry periods, such as the severe drought that plagued the northeastern United States during much of the 1960s.

## Drought Types

There are two types of drought—defined by the effects of the drought on man's activities. ***Agricultural drought*** occurs when rainfall is insufficient to maintain adequate growth and development of crops. At critical stages of crop development even a few weeks of dry weather can be disastrous, especially if accompanied by high temperatures. On the other hand, ***meteorological drought*** refers to a longer-term rainfall deficit which lowers ground-water tables, stream flow, and reservoir levels. Agricultural drought responds more rapidly to current weather conditions, while meteorological drought is slower to develop but also requires more persistent and frequent rainfall to bring to an end. Well-timed showers may supply sufficient moisture to the soil to maintain adequate plant growth, but may do little to replenish reservoirs and water tables. Thus a meteorological drought may persist, even though moisture conditions are sufficient for crop growth. By the same token, a few weeks of dry weather at critical times during the growing season may lead to agricultural drought conditions, but will not seriously affect stream flow or reservoir levels.

Agricultural drought involves four basic parameters—rainfall, evapotranspiration, initial soil moisture, and water requirements of plants. As long as the initial soil moisture at the start of the growing season and additional rainfall during the season are sufficient to satisfy the water needs of the plant and the water lost through evapotranspiration, then there will be no moisture stress on the plant. But if the soil moisture becomes deficient, then drought has begun. The higher the temperature, the greater will be the rate of evapotranspiration. Thus, drought conditions will develop more rapidly during a dry period if temperatures are above normal than if cooler weather prevails.

## Protection Against Drought

For the home gardener or farmer, the obvious protection against drought is irrigation. With most gardens or small farm plots irrigation is feasible, unless the drought is so severe that water use is restricted by the local water company, or your well is dangerously low. With large acreage, however, serious consideration must be given to the benefit/cost ratio before a decision to irrigate is made.

One protection that all growers can take, regardless of plot size, is to provide large amounts of organic matter in the soil. A rich, deep, well-balanced soil will have a much better water-holding capacity and will also encourage a large, vigorous root system. A vigorously growing root system is essential to reduce the effects of a dry period. By occupying

a larger and deeper volume of soil, the roots have access to a larger volume of water than do poorly growing shallow roots.

The use of mulch is also a primary method of protection against prolonged dry weather. Mulch not only minimizes evaporation losses from the soil, but by keeping the soil surface cooler it reduces the heat stress upon the plant as well. By keeping weed growth to a minimum, the mulch also aids in conserving water loss by reducing competition for available moisture. As mulch decays (assuming, of course, it isn't black plastic or stones), it also provides additional organic matter to the soil, further improving its water-holding capacity.

In addition, a good mulch will prevent the top layer of soil from getting "baked." Then, when rain does occur, much more water is absorbed by the mulched soil and is thus available for plant use, while less water runs off. Suitable organic mulches include grass clippings, hay, straw, shredded leaves, sawdust, wood shavings, seaweed, and peat moss. A layer or two of newspaper under the mulch will virtually assure a weed-free garden. A good discussion of the types and uses of mulch is found in *The Basic Book of Organic Gardening*.[1]

Another practice that helps reduce water loss from the soil is planting in narrow rows. With plants closer together, the foliage acts as a cover to reduce evaporation and maintain cooler soil surface temperatures, which also reduces evaporation demand. The less open soil, the less the evaporation. Note that the use of more narrow rows should be accompanied by somewhat greater spacing of plants within rows, so that the total number of plants remains about the same. Any significant increase in plant numbers within a given area will of course increase the demand for both water and nutrients. By careful spacing, however, the same number of plants can be made to produce a much more complete coverage of the soil surface. Not only will such an arrangement reduce water loss from evaporation, but it also provides maximum use of the solar energy striking the plot.

With a more complete foliage coverage there will also be less splash and runoff during rainfall. The more favorable soil moisture conditions caused by either mulching or narrow-row plantings will encourage the deeper, more vigorous root development which is desired as additional insurance against drought damage.

Another protection against drought losses is the use of plant varieties that have been bred to resist drought. You should check your seed catalogs or your local Extension Service to see if such varieties are available for the particular crop you wish to plant.

*For most garden vegetables, a total of about one inch of water per week is adequate.*

## Irrigation

If irrigation is required, how often should you water and how much water should you apply? The exact answer depends on what you are growing, the type and condition of your soil, the use of mulches, and general weather conditions over the past week or two. Your best advice probably can be obtained from your local county agent or agricultural experiment station. Some general guidelines follow:

If you are growing one, or only a few, specific crops, you should get information on the precise moisture needs of those crops. For example, it has been found that on most soils on which strawberries are grown, the plants require about 1½ inches of water each week from the period before blossoming through the end of harvest, and then about one inch per week the remainder of the growing season to ensure high yields the following year.

For the more general types of gardening where a wide variety of crops are grown, a good round figure of about one inch of water per week should be adequate. Obviously, this will vary depending on weather conditions, soil type, density of planting, mulching, etc. But for general guidance, the one-inch-per-week rule is quite useful and simple. I have found this rule-of-thumb to be adequate for my garden in south-central Pennsylvania. In fact, as the soil's organic matter content has been built-up and increased use made of a newspaper/grass-clipping mulch, the water requirement has been reduced and the need for irrigation even during dry periods has been lessened.

Remember, of course, that the required water can come from rainfall or irrigation, or a combination of both—it is not the amount to be applied in addition to rainfall. You should also be flexible. That is, if the rainfall over the past seven days has been only 0.25 inch, but showers are forecast within the next day or two, there is not much point in watering the garden.

Wait and see how much rain you receive. Even if none develops, your plants won't be hurt. Also, if the weather has been extremely wet, a week or two with little additional rainfall will actually be good for the garden. Excessively wet soils are as damaging to plants as are soils that are too dry. As strange as it may seem, overly wet soils can cause a reduction in the water taken up by the plants, because of the effect of a lack of oxygen on the physiological condition of the plant.

*Drip irrigation is being used here for a young peach tree near Fresno, California.*

Unfortunately, there isn't very much that the home gardener can do about excessively wet conditions. A good, well-drained soil, rich in organic matter, is probably your best protection. The use of tile drains, although effective, is usually feasible only in special situations. If poor drainage is a problem, check your local Soil Conservation Service office for professional advice.

For gardeners with soil moisture instruments, the actual moisture condition of the soil can be determined regularly, and water applied as needed. Irrigation is usually recommended when the available water in the root zone has been depleted by about 60 per cent. Enough water should be applied to bring the entire root zone (not just the surface layer) up to field capacity.*

Recently, the use of ***drip irrigation*** has been gaining popularity with home gardeners. Its primary advantage is a much more efficient use of irrigation water. With the typical sprinkler system, part of the water is evaporated in the air or from the plant foliage before it ever reaches the soil. With drip irrigation the water is slowly applied directly to the soil along the plant row, where it is most needed. An interesting discussion of drip irrigation appeared in the February 1977 issue of *Flower and Garden* magazine.[2]

For gardeners who have evaporation measuring equipment, a running account of the water balance of your garden can be kept. Start early in the spring when it is usually safe to assume that the soil is saturated (except in arid or semi-arid regions such as the southwestern United States). Thereafter, keep daily or weekly records of total rainfall, irrigation, and evaporation. Subtract the amount of water evaporated from the rainfall (plus any irrigation) and you will have a general value of the net gain or loss of moisture from your soil each week. The amount of evaporation you record won't necessarily be exactly the same as the amount actually lost through evapotranspiration in your garden, but it will permit you to determine if rainfall is keeping up with water loss or if supplemental irrigation is required.

Remember, too, that the effectiveness of rainfall depends not just on the amount, but also on its intensity and on the absorbing capacity of the soil. Here again, a good organic mulch on the garden will capture much of the rain that falls, even in a heavy shower, and will avoid excessive runoff and water loss.

---

*Field capacity represents the maximum amount of water that the soil can hold after the rapid downward flow of water due to gravity has ceased and all surface water has drained off.

## Drought Prediction

It is not possible to forecast drought conditions in advance, with any degree of accuracy, and the ability to do so in the foreseeable future is quite unlikely. There have been attempts to forecast major droughts, based on past drought "cycles." For example, there appears to be a tendency for severe droughts to occur roughly every 20 years in the central United States. The dust bowl days of the 1930s and the major drought of the 1950s were the most notable events in this cycle, but evidence for its existence has been traced back into the 19th Century. On this basis, there have been predictions of a disastrous drought in the U.S. heartland during the 1970s. So far (through the summer of 1978) a prolonged widespread drought has not developed, although various sections have had periodic problems with a lack of rainfall. While some of these "mini-droughts" have been sufficiently prolonged to result in significant crop losses to those involved, the overall crop-growing picture in the central United States has been generally quite favorable. Record or near-record grain crops were produced in the United States "breadbasket" in 1975, 1976, and 1977, so there is cautious hope that perhaps the 20-year drought cycle has either been broken, or at least seriously weakened. This does not preclude, however, the possibility of a serious drought situation developing in the next few years.

An interesting study of drought occurrences in the northeastern United States[3] suggests that a severe drought can be expected in that region about once in ten years on *average*, but not necessarily *every* ten years.

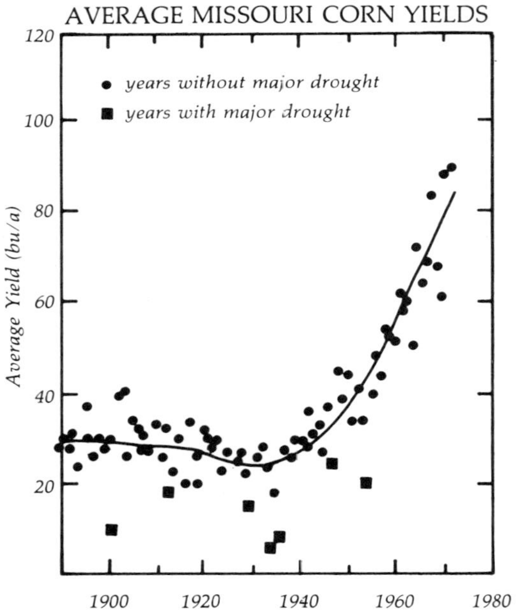


*This U.S. Department of Agriculture graph shows the relationship between drought and corn yields in Missouri for a period of more than 80 years.*

## Drought Severity

How do you measure drought? Unlike rainfall, evaporation or temperature, drought is not a uniquely defined parameter. There is no instrument to measure it and no universal agreement on exactly what conditions constitute a true drought. Until recently, only subjective descriptions of drought, such as mild, severe, and extreme, were used. In the 1960s, however, partly in response to the severe drought which gripped the northeastern United States, a drought index was developed by Wayne Palmer, a U.S. Weather Bureau meteorologist and one-time farmer. The Palmer Index attempts to classify drought conditions according to water needs in an area in relation to rainfall, soil moisture, water usage, and other factors which affect the water balance. This index is actually a measure of meteorological drought. A second index, the Crop Moisture Index (CMI) was subsequently developed by Palmer to measure the agricultural drought in a given area. National maps of the CMI are published weekly during the growing season in the national *Weekly Weather and Crop Bulletin* (see Appendix C), while Palmer Index values are published about once a month in the same publication.

*Wind and drought can cause severe conditions in the Great Plains, as seen in this photograph taken in Seward County, Kansas. In the winter of 1975-76, dry winds blew young wheat plants right out of the ground.*

## WIND

Wind damage to crops may be either local (as occurs with severe thunderstorms or tornadoes), or widespread (such as associated with typhoons and hurricanes). In the case of local severe storms, damage may be limited to only parts of a particular field, or may follow a strip cutting across many farms and gardens.

Most wind damage involves plants lodging (being bent over or broken down) and trees being uprooted or stripped of large branches. Whole fields may be laid low to the point where harvest of the crop is all but impossible. Orchard trees may be uprooted or, if winds are not that strong, fruit may be blown from the trees and damaged or destroyed. The problem of uprooted trees is particularly severe if the ground has been saturated by rain.

## Wind Speed Estimating Guide

| Natural Effects | Miles per Hour | Knots per Hour |
|---|---|---|
| *Smoke rises straight up. No movement of anything.* | *Under 1* | *Under 1* |
| *Direction of wind shown by smoke. Weather vane not affected. Tree leaves barely move.* | *1-3* | *1-3* |
| *Wind felt on face. Leaves rustle slightly. Weather vane will be affected if it is very free swinging.* | *4-7* | *4-6* |
| *Leaves and twigs are moving. Loose bits of paper are lifted and ruffled.* | *8-12* | *7-10* |
| *Small branches are moved. Loose dust and paper are blown along.* | *13-18* | *11-16* |
| *Larger branches and small trees sway. Dust clouds.* | *19-24* | *17-21* |
| *Larger branches in constant motion. Wind begins to whistle.* | *25-31* | *22-27* |
| *Whole trees are in motion. Walking is affected.* | *32-38* | *28-33* |
| *Twigs begin to break off. Walking becomes difficult.* | *39-46* | *34-40* |
| *Slight structural damage occurs. Gale winds.* | *47-54* | *41-47* |
| *Trees are uprooted and large limbs blown down. Heavy structural damage occurs.* | *55-63* | *48-55* |
| *Widespread damage. Very violent storm.* | *64-75* | *56-63* |
| *Severe destruction of almost anything. Hurricane.* | *Over 75* | *Over 63* |

Depending on the strength and duration of the high winds, the amount of damage can vary greatly. In some cases, plants may partially or fully recover. In small gardens, growers can often "repair" the damage using stakes and string, but on large farm plots this becomes impractical. Depending on the stage of development, fruit blown down in an orchard may be at least partially salvageable. Uprooted trees are a complete loss, except perhaps for the wood. Usually, areas affected directly by tornadoes are destroyed.

## Wind Protection

While it is virtually impossible to protect crops against severe winds, the use of windbreaks and shelter belts can provide some protection against the more frequent moderate winds associated with common thunderstorms and frontal passages. Such protection not only helps to prevent or minimize direct wind damage, but also affords some indirect protection against drought by reducing the amount of water loss due to evaporation. Such a factor may become important in times of moisture stress. Windbreaks also help to reduce soil loss due to wind erosion. The windbreaks may be natural, such as a strip of evergreen trees, or man-made structures designed to reduce the strength of the wind as it moves over the planted area.

The importance of a healthy and vigorous root system cannot be overemphasized. The roots of a plant are its "foundation," and just as the strength of a house's foundation determines its ability to withstand wind and flood, so a plant's root system largely determines its ability to withstand nature's onslaughts. Rich, healthy, fertile soils, well-watered, provide the medium necessary for a strong and vigorous root system. Thus, good gardening practices not only repay the grower with healthier, better-yielding plants, but also provide greater resistance against loss or damage due to various weather hazards.

*One of the best forms of protection from wind and snow is a windbreak, similar to this one in Clifford, North Dakota. The tree planting was started in 1948.*

# HAIL

Hail damage to agricultural crops in the United States amounts to approximately $700 million annually. Additionally, untold damage occurs to family gardens and ornamental plantings. Crop losses due to hail in other countries are not accurately known, but total world-wide losses certainly exceed $1 billion each year. Large hail may also cause injuries to animals, but specific data on such losses are unavailable.

In the United States, the greatest frequency of hail occurs in the west-central Great Plains, with an average of about eight to twelve hail days per year, while the lowest frequency is found in Florida and along the Gulf and Pacific Coasts. At first this may seem surprising, since the southeastern United States is one of the maximum thunderstorm areas in the world, and hail storms are usually, though not always, associated with thunderstorms. However, in southern regions the freezing level is considerably farther above the surface than it is in the higher elevations of the western Plains, and hence fewer of the storms reach levels where hail can form. In general, severe hailstorms are usually associated with temperate zone thunderstorms. Outside the United States, the major hail regions include Central Europe, northern India, and the Argentine Pampas.

*These peaches were ripped apart in a July 1975 hailstorm in Maryland.*

Hail damage is caused by the ripping and tearing action of the ice pellets as they beat down on plant leaves and stalks. Even where such injury does not kill the plants outright, they are weakened and thus more susceptible to disease and pest damage. Since hail stones vary from pea size or smaller to chunks of ice as big as softballs, the severity of damage also varies. In addition, both the duration of the storm and the density of hailstones (the number of stones hitting a particular area in a certain period of time) have an important effect on the extent of damage. Hailstorms are usually localized, seldom affecting an area of more than several square miles.

## Hailstorm Protection

There is, in truth, very little protection against hail damage. Plants alongside a building, hedge, or fence may receive partial protection, depending on wind currents during the storm, but only a small percentage of crop plants are so situated. Even the healthiest, strongest plants cannot withstand the savage beating of falling iceballs. And damage to exposed greenhouses can be almost complete.

Recovery of plants from hail damage de-

*Hail can reach large sizes, as shown in this photo of Joe and Jill VonNewham of Norborne, Missouri.*

pends on many factors, including severity of the storm, stage of growth of the plants, and subsequent disease and/or insect outbreaks. In severe cases, replanting (if feasible) is the only alternative to harvesting no crop at all. In other cases, where damage is partial, it pays to wait at least a few days to see how the injured plants will respond. Our own garden was literally "shredded" by a severe hailstorm June 1, 1976. Hail up to at least 1.5 inches in diameter fell for a period of 10 to 15 minutes, covering the ground to about an inch in depth. My initial feeling, following a survey of the garden immediately after the storm, was that most crops then in place were ruined. However, closer inspection the following day suggested a partial recovery might occur. A few destroyed plants were removed, but the remainder of the garden (which included broccoli, beans, corn, tomatoes, peppers, potatoes, carrots, peas, and beets) was left alone, and, with a few exceptions, final yields were unaffected. Had the storm occurred several weeks later, when plants were larger and more mature, permanent damage would have been greater.

About the only hope for the grower as far as reduction of hail damage is concerned, lies with research efforts into modifying storms to either prevent hail from forming or at least reduce the size and number of hailstones. Considerable work has been done along this line in recent years, especially in the United States and the Soviet Union. Some of the results are encouraging, but much work remains before the artificial reduction of damage from hailstorms becomes operational. Even then, the benefits to the home gardener will be minimal. Hail damage to gardens, although infrequent in most areas, will have to be endured by growers as an "act of God" over which they have little or no control. It's not always possible to foil mother nature!

# Part III:
# Using Local Weather Information

# 6
# Growing Degree Days —Better Than The Calendar

On most seed packets, there's a statement of the plant's required days to maturity. In most cases, the period to maturity is expressed in *calendar* days. A particular vegetable will be ready to harvest, or a particular flower ready to bloom, within so many days of planting or emergence. Thus, one variety of green peas matures within 60 days after planting while another will be harvestable after 68 days. Different varieties of sweet corn require from about 60 to more than 90 days before the first picking can be expected. Most, but not all, seeds purchased in bulk by farmers are also labeled in "days to maturity."

Knowing the number of days to maturity for each crop, the careful grower can plan a garden that will provide an extended period of harvest. By staggering planting dates and using several varieties with different maturity dates, the harvest period for a given crop can be significantly increased, or timed for maximum commercial advantage. However, if you keep close check on the time required for a particular crop to reach maturity you will likely find that the actual period does not necessarily agree with information on the seed package and will vary by several days from year to year. If you make two plantings of the same crop a week or two apart in the spring, you may notice the later planting requires fewer days to reach maturity than the earlier one, even though they both have the same maturity rating.

The basic reason for this is that growth and development of plants are not related directly to the number of calendar days, but depend on weather conditions—particularly air temperature. The calendar day method of labelling seeds is based on "normal" weather conditions that seldom occur, and does not take into consideration temperature variations from one area to another. The "days-to-maturity" rating represents a very broad average designed to cover wide regions of the country. Hence, the accuracy of the calendar-day method of estimating the maturity of a crop will depend directly on how "normal" the growing season is in any particular year. Even if temperatures are normal, growers in cooler northern zones may find their crops maturing slightly later than the date indicated by the calendar-day method of labelling. Southern growers, on the other hand, may have crops ready to harvest earlier than indicated.

## PLANT GROWTH AND TEMPERATURE

The relationship between plant growth and temperature is shown in the figure *(Plant Growth vs Temperature).* In general, the growth rate increases as the temperature increases, but the relationship is not the same at all temperatures. Hence the growth rate (solid) line in the figure is not straight but curved, sometimes referred to as an "S" curve. Each plant variety has a certain base temperature above which growth begins. At first, as the temperature rises above this base value, growth is slow but steadily increases until the temperature reaches a range where the growth rate increases the most rapidly. Further increases in temperature produce steadily smaller increases in the growth rate until finally an upper limit is reached, beyond which the effects of heat cause a decrease in the rate of growth and development.

As an example, the base temperature for dent (field) corn is about 50° F.; the most rapid increase in the growth rate occurs as the temperature rises from about 68° to 75°; and above about 86° the growth rate levels off and then begins to decrease. Similar limits apply to many other warm season crops, although more research is needed to define better the critical temperatures for many plants. For cool season crops, such as spring oats or potatoes, a base temperature of about 40° and an upper limit of about 80° are suggested.[1] This upper limit represents the maximum rate of growth and should not be confused with an upper *lethal* temperature which is higher and indicates the temperature at which serious injury or death of the plant will occur.

*Harvest time! Dates when vegetables will be ready to harvest may be determined by simply calculating the number of growing degree days from germination to maturity.*

## HEAT UNITS

A better indication, then, of the time required for a given crop to reach maturity is some measure of temperature that can be accumulated throughout the growing season. Such a concept was proposed as far back as 1735 by the French scientist Reaumur.* Since then, many techniques have been devised and tested for relating temperatures to crop growth through the use of what are known as heat units, or *degree days*—usually called *growing degree days* to differentiate them from heating degree days, which are used in winter to determine fuel requirements for heating buildings (see later section in this chapter). Although some of the methods are quite elaborate and complicated, it has been found that a simple growing degree day (GDD) formula, based on daily maximum and minimum temperatures is quite sufficient for most purposes and can be easily calculated and used by home gardeners and farmers. The grower can use either the daily maximum and minimum temperature as reported by the nearest weather station or, even better, use his own readings if he has a max/min thermometer.

*Rene Antoine Ferehault de Reaumur

### GDD Calculations

The basic formula for calculating growing degree days is:

$$\text{GDD} = \frac{\text{max.} + \text{min. temp.}}{2} - \text{base temp.}$$

PLANT GROWTH vs. TEMPERATURE

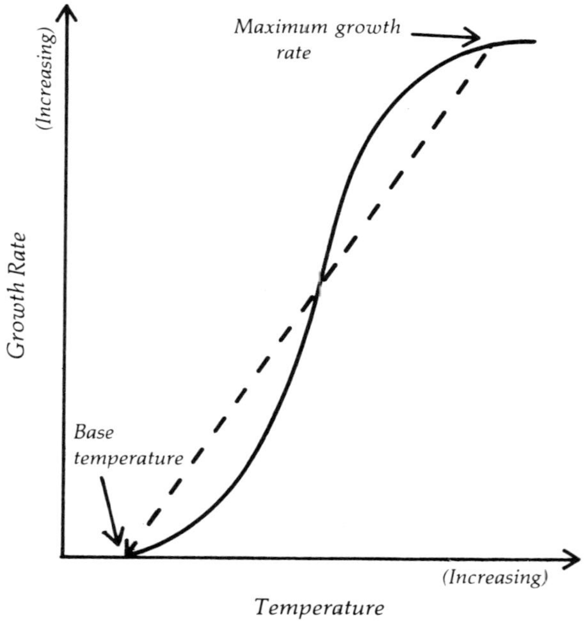


PERCENTAGE CORN TASSELED vs. GROWING DEGREE DAYS

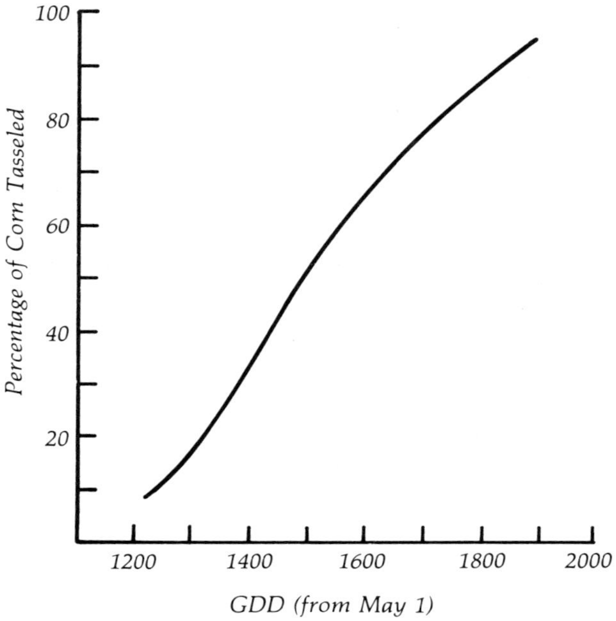


*Plant growth is relatively slow when temperatures are near the base level (for example, 40 for cool season crops such as peas); then the growth rate increases as temperatures increase; finally, growth again slows when temperatures are high.*

*These graphs, above and right, show the relationship between the rate of development of field corn and growing degree days. Above, the percentage of corn tasseled is related to growing degree days (from May 1).*

Thus, if the highest daytime temperature is 80°, the minimum is 60°, and the base temperature is 50°, there would be [(80+60)÷2]−50 = 70−50 = 20 GDD. In other words, subtracting the base temperature from the average temperature for the day equals the GDD for that day. If the mean temperature is less than 50° no GDD accumulate. Note that there are no negative growing degree days—that is, if the mean temperature for a particular day is 45°, the GDD for that day are *zero,* not minus 5. The reason for this is that while no plant development occurs below the base temperature, there is no negative development either. The plant growth rate simply remains stagnant. By calculating growing degree days for each day during the growing season and keeping a running total, the accumulated GDD for any particular period can be determined. It is best to begin the growing degree day calculations prior to planting time so you are sure to have records for your earliest plantings.

Even when using this method, the number of growing degree days needed to bring a given crop to maturity will vary from year to year. This is because, although air temperature is the single most important weather parameter affecting crop growth, other factors are involved, such as soil temperature and moisture, radiation (sunlight), and daylength. Also, this simplified method for calculating GDD assumes a direct (straight line) relationship between temperature and crop development (the dashed line in the figure, ***Plant Growth vs Temperature***). As already noted, the true temperature/growth relationship is represented by the

PERCENTAGE OF CORN MATURE
vs. GROWING DEGREE DAYS

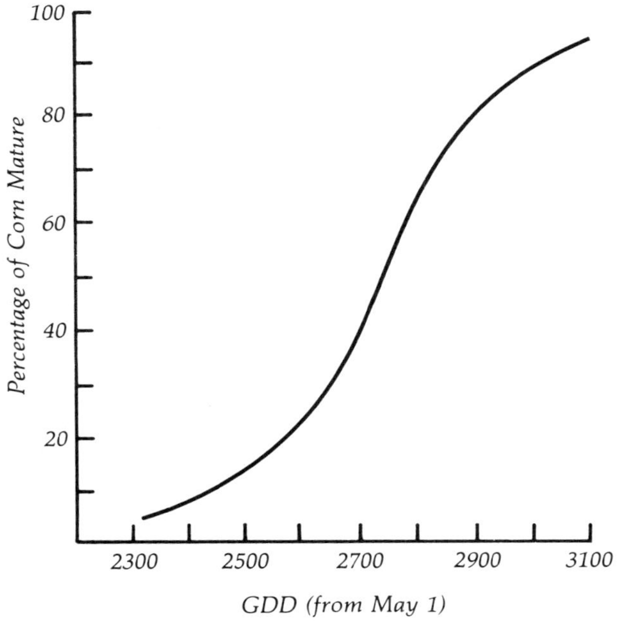


*Above, the percentage of mature corn is related to growing degree days (from May 1). The two graphs are based on data from the Corn Belt. The actual number of GDD required to develop tassels or reach maturity depends on the variety planted.*

curved line, but the straight line relationship is satisfactory and simpler to work with.

In any case, the use of growing degree days defines the period from planting to maturity much more accurately than does the calendar day method, which takes no account of weather conditions.

Examples of the relationship between the rate of development of field corn and growing degree days are shown in the figures, ***Percentage Corn Tasseled vs. Growing Degree Days*** and ***Percentage Corn Mature vs. Growing Degree Days.*** These graphs are based on data from the Corn Belt. The actual number of GDD required to develop tassels or to reach maturity depends on the variety planted.

## Modified GDD

Much formal research work on growing degree days has been done with field corn. One refinement in the GDD calculation technique produces somewhat better results than the standard GDD method, at least for field corn.[2] In this "modified" GDD method, the same basic formula is used, except that maximum temperatures above 86° are entered in the equation as 86°, and minimum temperatures below 50° are entered as 50°. Thus, for a day with a maximum temperature of 90° and a minimum of 65°, the calculations would be:

$$\text{Modified GDD} = \frac{86+65}{2} - 50, \text{ or } 25\ 1/2 \text{ GDD}$$

In practice, only whole numbers are used, so the actual value would be 26 GDD for that particular day. (The standard calculation method would show 28 GDD for that day).

The modified method is more accurate than the standard method because above 86° there is virtually no additional increase in the growth rate with an increase in temperature. These higher temperatures, then, do not contribute any additional growing "power." Also, while development may cease at temperatures below 50° there is no loss of previous growth and hence no loss of growing degree days.

This method of calculating growing degree days is used by some commercial seed corn producers to label their varieties of corn. It seems likely that, in the future, virtually all seed corn and perhaps other crops as well, will be labelled in GDD rather than by calendar days-to-maturity. Gardeners and farmers who have several years of temperature and growing degree day records for their location will be able to take advantage of this labelling method when and if it's commonly used.

While the modified GDD method represents a slight improvement over the standard calculation, for most growers the regular method is satisfactory. In fact, the critical lower and upper temperature limits are not as well defined for many crops as they have been for dent corn. Here again is an area for further research in which you, the home gardener, can try different methods and perhaps determine the best temperature limits for crop varieties in your region. The main point is: You need to be *consistent* and use the same method year after year to determine the average GDD needed for a given crop to reach maturity. At the same time you will be determining average daily, weekly, and seasonal GDD values for your location. A suggested technique for applying growing degree day methods to home gardens is presented later in this chapter.

## Solar Thermal Units

An advanced growing degree day concept has been developed by researchers at Montana's Agricultural Experiment Station[3], based on the idea that solar radiation, as well as air temperature, is a major factor in plant growth and development. They determined that mean daily temperatures become effective—upon plant development—when above 31°F. Once the daily average temperature is above this value, a combination of temperature and daily solar radiation directly control the growth and development rate of plants. This information has led to the concept of Solar Thermal Units (STU), which correlate quite well with plant development. Daily STU are calculated by multiplying the daily solar radiation (measured in $cal/cm^2$ by the mean daily temperature above 31°. If the mean daily temperature is 31° or less, no STU accumulate for that particular day. As an example, if the radiation for a given day is 400 langleys and the mean temperature is 61°, then:

$$STU=400\times(61-31)=400\times30=12{,}000$$

Lilac bloom when the accumulated total of STU (accumulations begin each year on January 1st) reaches about 380,000. STU requirements for blooming of these other plants and trees are:

| | |
|---|---|
| **Flowering dogwood** | 570,000 |
| **Common apple** | 380,000 |
| **Virginia strawberry** | 270,000 |
| **Alfalfa** | 820,000 |

Lilac and common apple tree blooms require the same number of Solar Thermal Units, and this means they should bloom about the same time each spring. Differences in varieties of either plant will, of course, cause some differences in bloom dates, but, for general purposes, the blooming date of one serves as a guide to the bloom date of the other. This relationship could be used to help select a site for an apple orchard, where a prime consideration is the susceptibility of the apple trees to a late spring frost. By noting the average bloom dates of lilacs in the area you can make an accurate estimate of the average bloom date of apple trees. Then, you can check local frost date records and determine the feasibility of establishing an apple orchard in that area.

The STU concept helps relate environmental conditions to plant development. One serious drawback to its widespread use, especially for home gardeners and farmers, is the lack of available local solar radiation data. Weather stations which record solar radiation information are few and far between and the data are seldom reported by popular media. While instruments to record solar radiation can be purchased (see Chapter 1) their high cost limits their availability to average homeowners.

LILAC/APPLE BLOOM
1965–1977

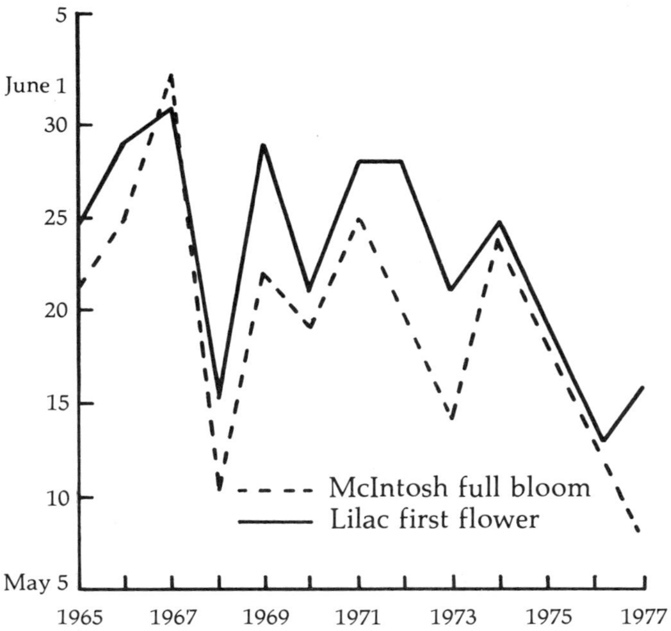


*A study by the University of Vermont Agricultural Experiment Station shows the close relationship between the dates of McIntosh apple full bloom and "Red Rothomagensis" lilac first flower. Being able to predict the apple bloom date could aid orchardists in preparing for the event.*

Nevertheless, future research into the STU/plant-growth relationship should provide useful information for growers—such as the lilac/apple-bloom example cited above.

## USE OF GROWING DEGREE DAYS

Growing degree days may be used to determine various stages of growth as well as the period to final maturity. From seeding to emergence, soil temperature is of prime importance and GDD are of minimal value, although air and soil temperatures are related closely. Once a plant emerges, however, air temperature becomes the dominant growth factor and a good correlation develops between GDD and crop development. Growing degree days can then be used to define such periods as emergence to bloom, bloom to fruiting, fruiting to harvest, bloom to harvest, or any other period of interest. The ability of growing degree days to adequately define a period of development will vary from crop to crop and from period to period for a given crop.

### Table 10: Growing Degree Day Calculations for Peas (BASE 40°F.)

| Crop Year | GDD Emergence to Bloom | GDD Bloom to Harvest | GDD Emergence to Harvest |
|---|---|---|---|
| 1 | 650 | 542 | 1192 |
| 2 | 717 | 529 | 1246 |
| 3 | 671 | 544 | 1215 |
| 4 | 699 | 503 | 1202 |
| 5 | 654 | 499 | 1153 |
| Average: | 678 | 523 | 1201 |

Table 10 presents an example of five crop-years of GDD calculations for a variety of peas, based on data I accumulated for our garden. The GDD were calculated using a base temperature of 40° F., which is typical for cool-season crops, and the standard GDD equation. With these data, once a planting of peas has bloomed, and knowing the average number of GDD for each day (or week), an accurate estimate of the date of first harvest can be made. Say, for example, that bloom occurs on May 25th and the average daily GDD (base 40°) is 20 for the last week in May, 22 for the first week in

June, 24 for the second week in June, and 26 for the third week in June. Then:

| | Accumulated GDD Since Bloom |
|---|---|
| **average GDD for May 26-31** (6 days×20 GDD/day) =120 | 120 |
| **average GDD for June 1-7** 7 days×22 GDD/day)=154 | 274 |
| **average GDD for June 8-14** 7 days×24 GDD/day)=168 | 442 |

Thus, by June 14, 442 GDD normally will have accumulated since May 25th, the date of bloom. Since 523 GDD are required, on average, between bloom and harvest (from Table 10), we need about 80 more GDD before the peas can be harvested. With an average of 26 GDD per day during the third week of June, the 523 total GDD should accumulate by about June 17th (another three days at 26 GDD/day= 78 GDD, plus the original 442=520 GDD). During the five-year period of record in this exam-

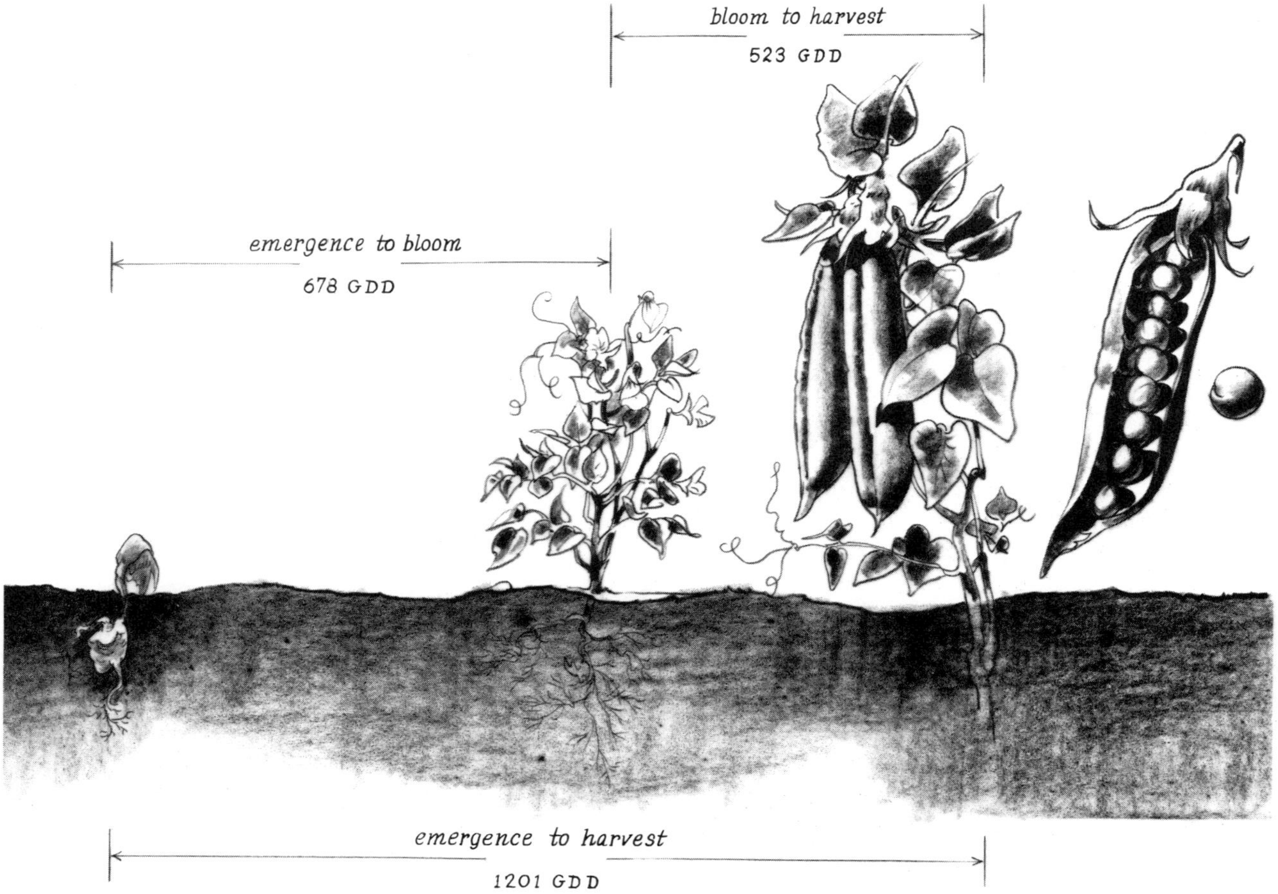


*Using growing degree days calculated from a simple formula, the author has been able to predict accurately the development of his pea crop. On an average, his peas are ready to harvest 523 growing degree days after they bloom. Over five years, the harvest never occurred more than about one day earlier or later than the date predicted.*

ple, the seasonal total of GDD from bloom to harvest ranged only about 20 GDD above or below the 523 average. This represents no more than one calendar day. Thus, if the peas bloom on May 25th, I can reasonably expect the first harvest of fresh peas to occur from June 16 to 18. By keeping track of the actual GDD each day, I could spot any significant differences from normal values and update my estimate of harvest time.

You can use the same technique to estimate the date of harvest for almost any crop, after you have accumulated several years of data. The date of harvest can be estimated from the date of planting or emergence, as well as from the date of bloom. Also, you can plan your planting dates so crops can be expected to reach maturity with a minimum risk of frost damage in the fall. And you can do so with more accuracy than simply using calendar dates.

The advantage of the GDD method over the use of straight calendar days is illustrated by Table 11. For the same set of data used in Table 10, the number of calendar days required between bloom and harvest are listed. While the harvest date occurred within plus or minus one day of the average using GDD (assuming an average daily GDD value of about 20), it varied from plus 3 to minus 5 days from the average using calendar days.

### Table 11: GDD vs Calendar Days

| Crop Year | GDD Bloom to Harvest | Calendar Days Bloom to Harvest |
|---|---|---|
| 1 | 542 | 22 |
| 2 | 529 | 20 |
| 3 | 544 | 19 |
| 4 | 503 | 14 |
| 5 | 499 | 21 |
| Average: | 523 | 19 |
| Variability: | +21 to ± 24 GDD (about ± 1 day) | +3 to −5 days |

The growing degree day method can be applied to fruit crops and ornamentals as well as vegetable crops. Some species will respond more reliably than others, but the basic technique remains valid. Each grower must determine the best GDD values for his region and plant varieties. While base temperatures also vary, some generalizations can be made to guide growers. For warm season vegetable crops a base temperature of 50° F. seems reasonable, while 40° can be used satisfactorily for cool season plants. One study of the use of heat units to forecast bloom dates of fruit trees suggested that for at least some of the common varieties of apple, pear, cherry, peach and apricot, a base temperature of about 40° to 44° would be suitable.[4] Base temperatures between 50° and 55° are appropriate for citrus fruits.

Much testing remains to be done to determine the base and critical temperatures for many plants, but interested growers can do this on their own, without waiting for word from the "experts." As little as five years of data can yield approximate values, which can be adjusted as you obtain additional data in later years. Whether you select a base temperature of, say, 40° or 44° is not critical. It is important to select one figure and use it consistently from year to year.

# HOME GARDENING AND GROWING DEGREE DAYS

Keeping track of growing degree days is quite simple, and the only instrument required is an inexpensive maximum/minimum thermometer (see Chapter 1). Many growers will find that monitoring GDD can be profitable: It helps time the maturity of crops to coincide with prime selling periods; and helps to avoid fall frost damage caused by selecting crops that take too long to mature.

## Growing Degree Day Record

Two types of records are required, and one important item to obtain is a good record book.

First, you will need to keep a record of daily maximum, minimum, and average (mean) temperature. These values should routinely be recorded on your monthly weather record, as described in Chapter 1.

The growing degree day data can be kept as part of your weather record or as a separate record sheet. A sample of a partial GDD record is shown in Table 12. The mean temperature is simply transferred from your daily weather record.

It is best to keep track of at least two different GDD accumulations—one for cool season crops and one for warm season plants. As mentioned earlier, a base temperature of 50° is generally used for warm season varieties, while 40° is considered good for cool season crops. Hence, the two columns in Table 12 for $GDD_{40}$ (base temperature of 40°) and $GDD_{50}$ (base temperature of 50°). You can use as many different base temperatures as you like, and by doing so may find that certain crops respond best to base temperatures other than either 40° or 50°.

**Table 12: Growing Degree Day Record**

Station ______ Month ____ Year ____

| Date | Mean Temp. | $GDD_{40}$ | Accumulated $GDD_{40}$ | $GDD_{40}$ | Accumulated $GDD_{50}$ |
|---|---|---|---|---|---|
| 1 | 45 | 5 | 5 | 0 | 0 |
| 2 | 51 | 11 | 16 | 1 | 1 |
| 3 | 57 | 17 | 33 | 7 | 8 |
| 4 | 48 | 8 | 41 | 0 | 8 |
| 5 | 60 | 20 | 61 | 10 | 18 |

You should keep a record of the GDD for each day, as well as a seasonal accumulation. As a general rule, begin the accumulation on the first day of the month in which you expect to do your first planting. Thus, if you usually plant your first crops during April, begin your GDD accumulation on April 1st. It is *extremely important to always use the same starting date each year,* so once you have selected a date—stick to it! Continue your accumulation from one month to the next, right through the growing season.

## Crop Record (Phenological)

The second type of record you need to keep includes the planting dates of each of your crops (for annual garden crops) and the dates of subsequent stages of development such as emergence, first bloom, first fruit, and maturity. Such records are often kept by gardeners anyway, to aid in planning the following year's garden.

For fruit trees and bushes, and for ornamental plants, the first stage recorded would probably be first leaf, followed by first bloom. Subsequent stages which should be noted (and their dates recorded) may include full bloom, end of bloom, and first fruit, depending on the characteristics of the specific plant.

To maintain such phenological records (see Chapter 2) you must make frequent checks of your garden and landscape areas—preferably daily. Such observations have a side benefit: You will quickly spot any disease or insect problem which might otherwise have gone undetected.

Your crop record sheet, then, should provide listings for dates of planting, emergence, first bloom, first fruit, and maturity. Not all columns will be filled in for all crops and other stages may also be added, such as leaf coloration and/or leaf fall in the autumn for certain trees or shrubs. There are no absolute rules. You can decide what you want to record as you go along. The primary stages shown above, however, are the most important ones and should be noted each growing season. The date of maturity would be the date when the crop is ready to harvest, even though you may delay the actual harvest a few days for one reason or another. Part of my phenological record for 1976 is shown in Table 13.

## Table 13: Crop Record (Phenological)

Station Camp Hill, PA    Year 1976

| Crop or Plant | Planting | Emergence | First Leaf | First Bloom | First Fruit | Harvest |
|---|---|---|---|---|---|---|
| Strawberries | — | — | — | April 19 | May 5 | May 24 |
| Peas | April 10 | April 19 | — | May 24 | June 6 | June 13 |
| Lettuce (loose-leaf) | April 10 | April 19 | — | — | — | May 3 |
| Carrots | April 13 | April 20 | — | — | — | June 30 |
| Sweet corn | May 15 | May 24 | July 4 (tassel) | July 12 (silk) | — | July 31 |
| Apple tree | — | — | — | April 13 | May 3 | July 29 |
| Lilac bush | — | — | March 12 | April 19 | — | — |
| Forsythia | — | — | — | March 21 | — | — |

*Records of the stages of plant development, such as first bloom, first fruit and maturity, are often kept by gardeners to aid in planning the following year's garden.*

## Combined Records

The final step then is to combine the growing degree day data with the crop phenological records. For each date listed in the crop records, note the seasonally accumulated GDD.

To determine the number of GDD required to advance the crop from one stage to the next, calculate the difference in GDD between the two dates. For example, one planting of peas in our garden emerged on April 24, 1975, blossomed on May 27, and was ready to pick on June 16. The seasonally accumulated growing degree days on each of these dates were:

| Date | $GDD_{40}$ Sum |
|---|---|
| April 24 | 157 |
| May 27 | 874 |
| June 16 | 1408 |

Thus, the GDD accumulated from emergence to blossoming were 717 (874-157), while 534 (1408-874) GDD accumulated between the time of the first blossom and maturity.

Similar calculations can be made for other crops. By collecting this type of data for several years, a definite pattern emerges as to the number of growing degree days required for each crop variety to advance from one stage of development to the next. If you change varieties you should begin a new record, since different varieties may have different GDD requirements. For this reason, it is a good idea to select one or more favorite varieties of a given crop and stick to those for determining the GDD/crop-development relationship.

Once you have established the GDD requirements for a particular crop, you can use your data to select planting dates that will produce a harvestable crop within a desired period. Thus, you can plant an early crop (taking into account local frost dangers) to mature at a favorable marketing time, or a late-season crop to reach maturity before the first likely frost in the fall.

Keeping weather and crop records might keep you one step ahead of neighbors or commercial competitors.

# HEATING DEGREE DAYS

Although not related to agriculture directly, the subject of heating degree days (HDD) follows naturally from a discussion of growing degree days. Probably more people are familiar with heating degree days than with growing degree days, especially since the start of the energy crisis, and the resulting concern over fuel conservation.

The basic method for calculating heating degree days is the same as for calculating growing degree days, in that the first step is to determine the average temperature for each day. This is done by adding the maximum and minimum temperatures together and dividing by two.

It has been determined that, in general, some heating is required in homes and office buildings if the outside air temperature averages below 65° F. on any given day. The amount of heat required (and, hence, fuel needed) is directly proportional to the number of degrees below 65°. Thus, a day with an average temperature of 45° accumulates 20 HDD and requires about twice as much fuel to maintain a given inside temperature as a day with a mean temperature of 55°, which accumulates only 10 HDD. The general formula is:

$$\text{HDD} = 65 - \frac{\text{max} + \text{min.}}{2}$$

No heating degree days accumulate if the

## HEATING DEGREE DAYS

*This map shows normal total heating degree days. Heating degree days are the number of degrees the daily average temperature is below 65°F.*

daily mean temperature is 65° or higher. By keeping track of HDD and fuel consumption you can determine the approximate number of HDD it takes to consume one gallon of fuel oil, one ton of coal, or one cubic foot of gas. For example, if 160 gallons of fuel oil are required to fill your tank, and 800 HDD have accumulated since your last oil delivery, you are using one gallon of oil for every 5 HDD (800 HDD/160 gal.=5 HDD per gallon). Oil companies use this method to schedule automatic delivery of oil to their customers. For those who order their own fuel, keeping a record of the HDD will prove to be a good guide to fuel requirements.

In addition, by keeping careful record of your fuel consumption in terms of heating degree days, you can spot deficiencies in your heating unit and arrange to have it inspected and serviced to restore peak performance. Also, more efficient fuel use due to energy-saving improvements, such as storm windows and insulation, will be readily apparent. This type of fuel efficiency analysis can be applied to greenhouses as well as to homes and other buildings.

The official heating-degree-day season begins July 1 and ends June 30.

# 7
# Understanding and Using Weather Forecasts

When you pick up a newspaper, usually the item at the very top corner of the front page is the weather forecast—often in an abbreviated form. A more detailed forecast, perhaps accompanied by a weather map and local weather statistics, is located elsewhere in the paper. The final item on most local TV and radio news programs is the weather forecast; exceptionally severe or unusual weather conditions may be a lead news story. Weather forecasts are repeated throughout the day on many radio stations.

In various forms, then, the local weather forecast is probably the most readily available and most frequently repeated item of information presented to the public. Most people hear or read the latest weather forecast several times each day, and yet there is probably no other single piece of information so misunderstood or misinterpreted by the very people it is supposed to help.

Part of the problem involves the manner in which the forecast is communicated to the public—a problem discussed later in this chapter. Also, forecasters are not always adept at wording the forecast so it says what they intend. At the same time, people hearing the forecast on radio or TV may not always be listening closely. A few key words, such as "rain," or "windy," may stick in a listener's mind, but his ability to recall other parts of the forecast is uncertain. Is the rain expected today or tomorrow? Was that frost warning for tonight or tomorrow night?

A better understanding of weather forecasts would obviously benefit many people. And no group is likely to benefit more than home gardeners and farmers.

## WHAT DOES IT MEAN?

"Mostly sunny and warm today, but with a slight chance of an afternoon or evening thundershower. High in the upper 80s. Fair and mild tonight, low in the upper 60s. Partly cloudy and more humid tomorrow with a chance of showers or thundershowers, high in the mid- to upper-80s. Probability of precipitation 20 per cent today, 10 per cent tonight, and 50 per cent tomorrow. Winds light and variable today and tonight, southwesterly 15 to 20 miles per hour tomorrow."

The above is a typical summertime weather forecast in many sections of the United States. Exactly what information does it convey? In particular, of what use is it to the home gardener or farmer who has to make decisions based on the expected weather? For the farmer with freshly cut hay, the forecast will influence his decision whether to let the hay remain in the field another day to dry completely, or to bring it in a day early to avoid a soaking rain and perhaps rotting. For the home gardener the forecast may influence a simple decision such as, "should the garden be watered or can it be left alone another day?" A host of other decisions by the grower may depend on expected weather conditions—decisions concerning sowing, spraying, cultivating, fertilizing, irrigating, harvesting, etc. In many cases, the results of these decisions can be the difference between failure or success of a crop. We all know weather forecasts are far from perfect, but they do represent the best "guestimate" of trained meteorologists, and the weather-wise gardener/farmer will make the most of them.

Take, for example, the forecast given at the beginning of this section—a forecast that is fairly typical for many areas of the country during the late spring and summer. Let's look at it in detail and see what useful information can be extracted, even if it turns out to be not entirely correct.

The basic item for "today" is that it will be "mostly sunny and warm." Not cloudy or even partly cloudy, but generally sunny. In addition to indicating the general sky conditions, this also suggests that: With temperatures in the upper 80s, a person working outside in the open fields will find it quite warm, if not downright hot; and, in the absence of any statement indicating high humidity, general drying conditions will probably be pretty good and any moisture on the soil or vegetation early in the morning should evaporate quickly. Note that if humid conditions were forecast, drying would be slower.

The additional statement concerning a slight

*A meteorologist prepares to release a weather balloon at Hallett Station in the Antarctic.*

chance of a thundershower is not, as some might think, simply thrown in by forecasters to cover themselves in case one should develop unexpectedly. It reflects actual atmospheric conditions. That is, while at any given point there is little likelihood of a shower, there may be a few showers scattered over a wide area. Most areas will *not* receive rain, and in fact none may fall anywhere in the forecast zone, but the weather conditions are such that a small probability (in this case 20 per cent) does exist for a local shower at some time late in the afternoon or evening. It would be nice if the showers could be pinpointed more accurately, but weather forecasting simply is not that advanced, and probably will not be in the foreseeable future. This is not meant as a criticism of meteorologists, but simply a factual statement about the complexity of the earth's atmosphere.

The sample forecast for "tonight" is straightforward; it implies that precipitation is unlikely. For "tomorrow" an increasing likelihood of rainfall is indicated, directly or indirectly, in several ways. First, the sky condition is forecast to be partly cloudy. In other words, tomorrow will likely be more cloudy than today. The forecast of higher humidity indicates poorer drying conditions, and indirectly suggests a greater possibility of showers. It also suggests strongly that outdoor working conditions will be less comfortable. The likelihood of showers has gone from "slight chance" today to "chance" tomorrow. This may seem trivial to casual listeners, but in the meteorologist's vocabulary it's significant. Finally, the probabilities themselves increase from 20 per cent today to 50 per cent tomorrow.

The final information given in our sample forecast, on wind speed and direction, is important to spraying or dusting operations. "Light and variable" indicates wind speeds of under 10 miles per hour, with the direction changing from time to time and not blowing persistently from one direction. Windier conditions are expected the following day, and the wind will be persistent from a southwesterly direction. Note that directions given in the forecast are those *from* which the wind blows. Our sample forecast thus suggests that spraying would best be carried out today, rather than waiting for tomorrow.

# PRECIPITATION PROBABILITIES

The true meaning of precipitation probability is often misunderstood. The probability given in the forecast is the probability that measureable precipitation (at least 0.01 inch or more) will fall at any given point within the forecast area. It is the result, actually, of two probabilities—the probability of rain occurring at least somewhere in the forecast area, and the percentage of the area likely to be affected. For example, suppose there is an 80 per cent probability that a cold front accompanied by showers will affect a region during a certain period. Suppose further, that the showers are scattered and only about half (50 per cent) of the area over which the front moves will actually have rain. The true probability of rain at any given point in the region is then $.80 \times .50$, or .40 (40 per cent). And remember, if there is a 40 per cent chance of rainfall, there is a 60 per cent chance of *no* precipitation. Another way to interpret a 40 per cent probability is that for every 10 similar weather situations, precipitation can be expected, on the average, about four times, while no rain is expected for the remaining six.

Mention of precipitation in the forecast is usually made only if the probability is 20 to 30 per cent or higher. For a 20 per cent probability, the forecast may omit the mention of precipitation or use the term "slight" or "small" chance. For a 30 to 50 per cent probability, the word "chance" is usually used in the forecast, while "likely" refers to a precipitation probability of 60 to 80 per cent. When the risk of precipitation is above about 80 per cent no qualifying term is used—simply a statement that rain or snow is expected.

While the precipitation probability percentages may not always be accurate, the *trend* from one forecast period to the next can be significant. In our example, the important information in this respect is that the probability increases from 20 per cent today to 50 per cent tomorrow—a good indication that atmospheric conditions are changing in a way that increases the likelihood of showers. Whether the percentages in the sample forecast are accurate or not (and in fact, of course, it either does or doesn't rain at a given location—there is no such thing as a 50 per cent rain), the trend in the probabilities suggests that if you have an outside activity that must be accomplished either today or tomorrow, you would be more likely to get it done today than 24 hours later. Basically, this is a simple interpretation, but one often overlooked by casual listeners or readers of weather forecasts.

# WATCHES AND WARNINGS

There often seems to be much confusion over two types of special weather messages—watches and warnings. These terms are quite important in understanding bulletins and special forecasts involving severe weather conditions.

A ***watch*** indicates conditions are favorable for a certain weather event to occur in a specified area during a specified period of time. Thus, a severe thunderstorm watch, for example, is a message to the general public indicating severe storms *may* be in the area during a particular time period, and that persons should plan accordingly. If a person could be affected directly or indirectly by such storms, then he should be alert to their *possible* occurrence. A watch *does not mean* the weather event will occur, or even that it is likely to occur. It means, rather, that there is a possibility of it occurring and that the public should be alert to that possibility. The farmer out in the field should be extra alert to changing sky conditions and be prepared to seek shelter if a storm materializes. Picnickers or hikers may not want to stray too far from their cars.

A ***warning***, on the other hand, indicates a certain weather event is either occurring or is imminent. For example, a severe thunderstorm warning means severe thunderstorms, possibly with strong winds and/or hail, have been reported nearby and are headed for the warning area. A warning (which could also be for such events as a tornado, flash flood, and heavy snow) leaves little doubt that the event *will occur* and proper precautions should be taken immediately.

Remember these two terms and their meaning. They could save your life or help you prevent or reduce damage to property.

*Flood conditions, such as these in Pennsylvania, can arise quickly. The National Weather Service issues* warnings *when flood conditions are about to occur.*

# WHO MAKES THE FORECAST?

While most forecasts are made by the National Weather Service (formerly the U.S. Weather Bureau), an increasing number of private meteorologists are issuing public weather forecasts. Usually they contract with radio or television stations or local newspapers to provide routine forecasts for local areas. Their major selling point is that they can provide a forecast tailored specifically for a small area (or a specific activity, such as agriculture or skiing), while the National Weather Service (NWS) must generalize forecasts to cover much larger regions and serve a more general audience.

Also, federal forecasters are restricted in their freedom to deviate from "guidance" material provided by a national weather center in Washington, D.C. Automation, computerization, and other forms of "progress" have invaded the forecasting field, but do not necessarily mean a better product for the public. How much this increasing centralization of NWS forecasts affects the accuracy of local predictions is debatable, but the fact remains: More forecasts by private meteorologists are being heard in many areas.

In areas where both NWS and private meteorologists issue weather forecasts, the listener or reader could receive conflicting reports. Much of the time, however, there is little basic difference between them, although the wording may vary. The greatest differences usually occur at times of inclement weather, when the exact timing, the total amount, and the type of precipitation (rain or snow) are in question. Both private and government weathermen base their forecasts on the same raw weather data and computer information, so differences in forecasts depend mainly on interpretations of those data. Since no individual has yet been found who is able to correctly predict weather at all times (although there may be a few who would have you believe otherwise), over the long run, National Weather Service forecasts and private forecasters are probably equally accurate. On any given storm, one might prove more correct than the other, or they might both "blow it" in different ways.

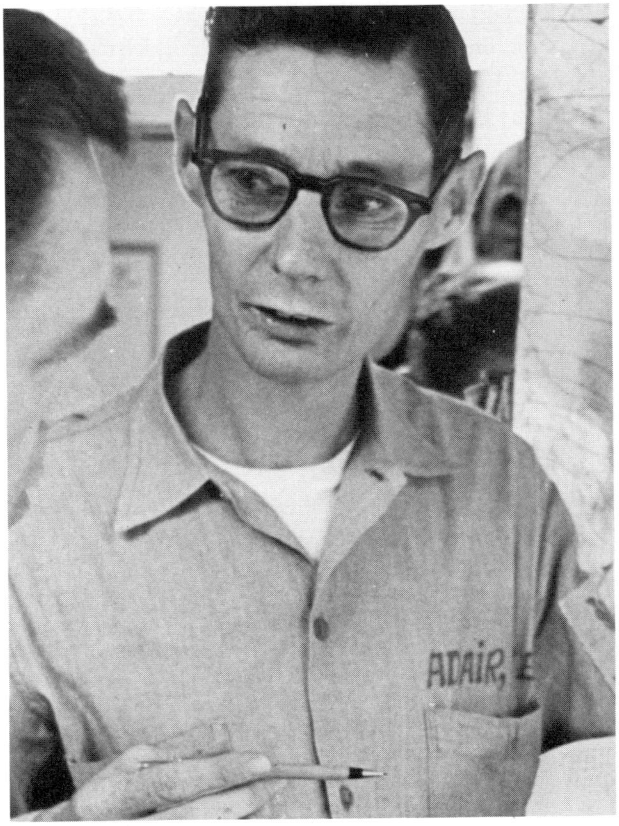


*The National Weather Service is the country's major source of weather information. Armed forces personnel and private meteorologists also issue forecasts.*

Generally, National Weather Service forecasts are issued about 5 A.M. and 11 A.M., and again about 5 P.M. The morning forecasts cover the current day, "tonight," and "tomorrow."

The evening prediction covers "tonight," "tomorrow," "tomorrow night," and the following (second) day. Thus, the Monday evening forecast, for example, would cover the period through Wednesday. If conditions change significantly, an additional forecast may be issued about 10 or 11 P.M. Updated forecasts are released whenever the weather changes suddenly and bulletins are issued to warn of possible severe weather conditions.

Long-range outlooks, covering the third through the fifth days, are issued each evening. For example, on Monday evening an outlook is issued for Thursday through Saturday. These outlooks are less detailed than regular forecasts. Outlooks covering temperature and precipitation trends for days six through ten are issued three times a week. A 30-day outlook is released twice a month. These monthly outlooks indicate whether temperature and precipitation are expected to be above or below normal over a given area during a particular 30-day period. No attempt is made to indicate when wet/dry or warm/cold periods may occur during the month. Seasonal temperature outlooks covering a three-month period have also been made available to the public. These are issued four times a year and do not give any information on precipitation amounts.

Private meteorologists may vary the timing of their forecasts as well as the length of the forecast period, since they are not tied to specific formats and regulations as is the National Weather Service. Many tend to make more frequent forecast updates than does the NWS. Some private forecasters extend their predictions an extra day or so beyond the NWS forecasts, but may not issue the equivalent of the three to five-day, or six to ten-day outlooks. Other private meteorologists have become known for their long-range forecasts. Unfortunately, forecast accuracy decreases rapidly after about two to three days—for either private or NWS predictions. In most cases, an accurate forecast for the next 24 to 48 hours is of greatest benefit to gardeners and farmers. Obviously, detailed long-range forecasts covering several weeks or months would be useful to agriculture, but reliable forecasts of this type are not likely to be available in the foreseeable future.

In some sections, specialized agricultural weather forecasts are issued by the National Weather Service. These emphasize aspects of the weather (for example, surface winds for spraying, or drying conditions for hay-making) that interest farmers. To find out if your region is covered by this agricultural weather service, contact the nearest NWS office or your county agent.

*Jack Hummel, meteorologist at the National Weather Service office in Burlington, Vermont, checks the station's latest weather data.*

# GETTING THE CURRENT FORECAST

A major weather forecasting problem is getting accurate and timely information to the users through the mass media. Forecasts in daily newspapers are hours old by the time they reach the public. In most cases, the published forecast has been superseded by a later one. Forecasts on radio or TV *should* be up-to-date, and in many cases they are. Unfortunately, there are also stations that use outdated forecasts, some that are up to 10 to 12 hours old. During rapidly changing weather, they are useless—except to confuse the public.

Another problem occurs when a radio or TV announcer, or "DJ", gives his own version of the forecast. "Partly cloudy" or "some cloudiness" becomes simply "cloudy"; "slight chance of some light snow late this afternoon" may become "snow today." The whole meaning of a forecast can be changed by varying or omitting a word or two here and there. Also, in some cases, a forecast for the wrong area may be given.

In some areas, weather offices have automated telephone lines which provide the latest forecast continuously. Often during periods of bad or threatening weather, however, and especially on weekends, these lines become overloaded. For most people, then, the radio or television is the best source for weather information. To be sure you're getting the most reliable and up-to-date forecast, pick a station you know is concerned with providing accurate and current weather information and stick with it. After listening to a station regularly for a short period you can tell whether it provides reliable weather data. If your local station seems disinterested in weather information, call or write the management. Let them know you would like accurate and up-to-date forecasts for your area. On the other hand, if a station does provide good weather information let the manager know the forecasts are appreciated. Better still—let the sponsors know!

In some sections, continuous weather information, provided by the National Weather Service, is available on a special radio frequency. The number of areas with this service (known as the NOAA Weather Radio) is increasing, and your area may be covered. Check with your nearest weather station before purchasing any of the popular special weather-band radio receivers. Where available, these continuous weather broadcasts are your best possible source of up-to-date weather information. Finally, some radio stations, particularly in a city where a National Weather Service office is located, often feature broadcasts direct from the weather office. Again, check with the local weather service office to see if these direct broadcasts are available in your area. (See Appendix C for a 1978 listing of NOAA Weather Radio Transmitters.)

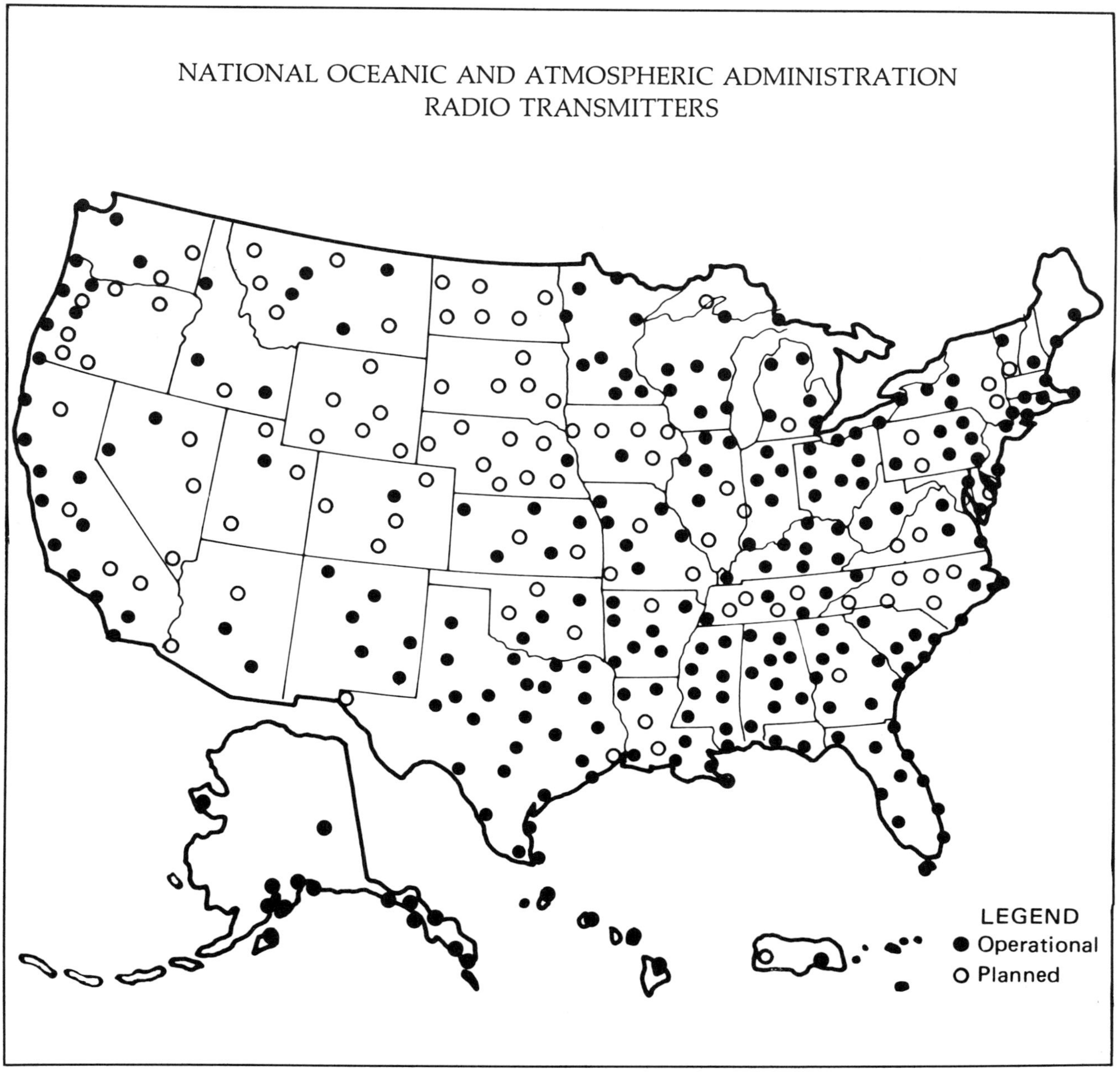


*The solid black dots indicate the location of NOAA Weather Radio Transmitters as of December 1, 1978. The white circles indicate those planned. These stations provide continuous weather information on assigned frequencies. For a complete listing of stations and frequencies, see Appendix C.*

# 8
# Climatology as a Guide

Many industries and businesses are directly or indirectly affected by weather and climatic conditions. None is as dependent upon weather for day-to-day operations as is agriculture. And no industry or business could benefit as greatly as agriculture from reliable long-range weather forecasts. Accurate predictions of growing season weather several months in advance would allow growers to set planting dates, determine fertilizer and irrigation requirements, be alerted to potential insect and disease problems, make a good estimate of quantity and quality of yield, and plan harvest dates—all before seeds were in the ground. Price variations in agricultural commodities would be stabilized as much of the uncertainty about future supplies would be removed. If factual knowledge about climate and possible climatic changes in future years were available, rational decisions could be made concerning the need for large-scale changes in world-wide growing patterns. For mankind in general, the threat of food shortages would diminish greatly, and, for growers, agriculture would become much less of an economic risk.

Unfortunately, forecasts that predict weather accurately, several months or a season in advance, are only a hopeful gleam in some meteorologists' eyes, and longer range predictions of climate are little more than science fiction.

This does not mean that information on future weather and climatic conditions is unavailable. While accurate daily or weekly forecasts, predicting conditions months in advance, aren't possible at present, information is available that allows alert growers to make intelligent and useful estimates of future climatic conditions and minimize the risk of loss due to adverse weather. The field of science providing this information is climatology.

Confusion often arises about the difference between weather and climate, and various wordy and sophisticated definitions have been devised to explain and clarify the difference. For the layman, it is quite sufficient and correct to say that weather is defined as the current day-by-day atmospheric conditions, while climate refers not only to "average" weather conditions over a long period, but also to the *variability* and *extremes* of these long-term conditions.

Thus, on any given day, or over a period of days, the weather in a particular area may include a wide range of events, but the climate of that region remains relatively constant over long periods. The growing season climate of a certain region, for example, is defined by the average or "normal" temperature, rainfall, sunshine, etc., and by the usual variations and extremes of these parameters over a period of years. Thus, one summer may be cool and moist, the next hot and dry, but the general climate of the area has not changed, because similar cool and wet, or hot and dry summers have occurred occasionally in past years and are a part of the overall climate.

The branch of science dealing with relatively short-term changes in the atmosphere, and which includes the familiar daily weather forecast, is known as meteorology, while the study of long-term atmospheric conditions is called climatology. (Actually, the distinction between the two fields is not that sharply defined and climatology is sometimes considered as a specialized branch of meteorology.) It is the field of climatology which can provide the farmer or home gardener with quite useful information on the likelihood of future weather conditions—information you can use in planning your activities one or more seasons in advance.

The climatologist, by carefully applying certain statistical methods in the analysis of past ("historical") weather data, can estimate the probability of future weather events. This information is not a specific "forecast" for a particular day or season, but rather an indication of the chance that a grower has of encountering a particular weather condition during a given period.

## FROST PROBABILITIES

Two of the most common examples of climatological data are the average dates of the last killing frost in spring and the first killing frost in fall. These dates, along with the average growing season rainfall, are probably the most widely used pieces of climatological information in agriculture. Many farmers and home gardeners rely on frost, or freeze, data to guide their planting and harvesting. But, these data are often misunderstood. Usually the dates referred to are the average dates of the last 32° F. or lower temperature in the spring and the first 32° or lower temperature in the fall. Some older frost data were based on observations of actual frost (hoarfrost) in the field. However, since frost can form at shelter temperatures above 32°, while at other times frost may not be

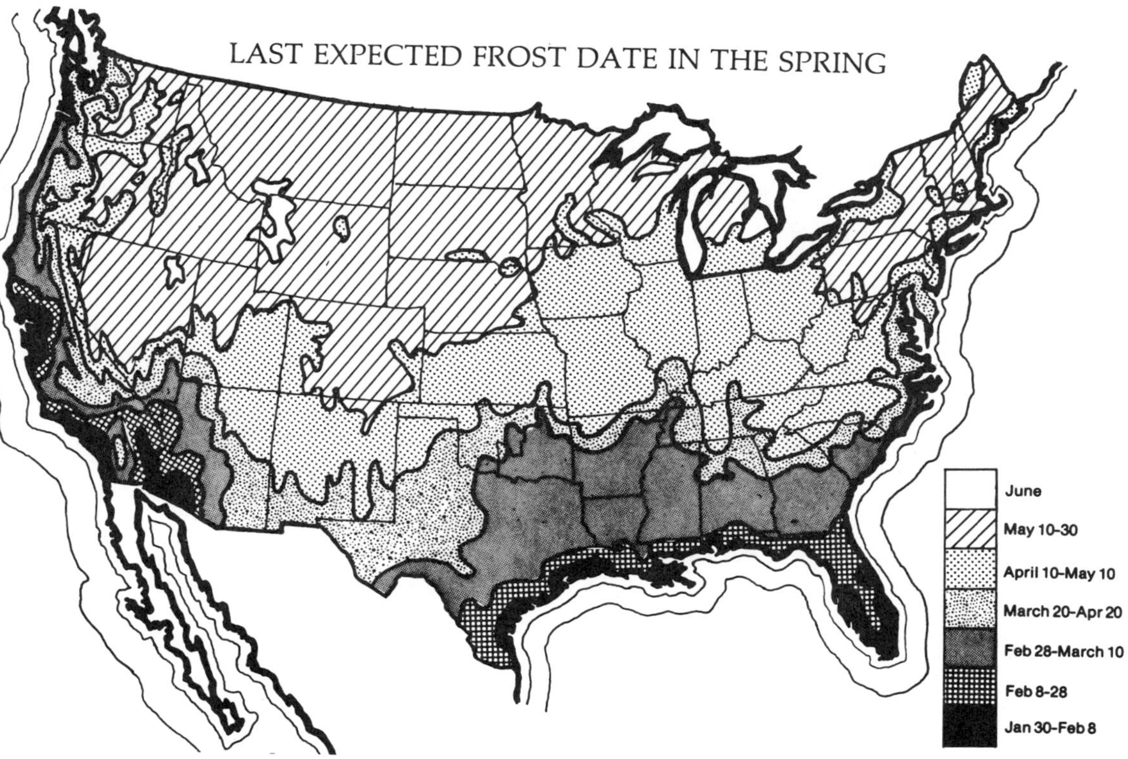


*This map shows, in general terms, when the last day of frost or 32° temperatures can be expected in the spring. The map should be only used as a general guide. Local temperature variations, due to terrain or other factors, can be significant.*

visible even though freezing temperatures occur, the use of 32° as the critical shelter temperature has become standard in determining freeze or frost dates. (The temperature at soil or crop surfaces may be significantly lower than the official temperature measured in a weather shelter. Thus, frost may form with shelter temperatures above 32°. On the other hand, if the air is quite dry, or if skies are mostly cloudy and/or winds are brisk, frost is not usually observed even though temperatures may be well below freezing. See Chapter 5 for a more complete discussion of frost.)

Often there is confusion as to exactly what these freeze dates signify. The *average date* of the last 32° in spring simply means that, over a long period of time, in about half of the years the last 32° or lower reading will occur before that date and in about half of the years it will occur after that date. In other words, if the average date of the last spring freeze in your area is April 15, there is still a 50 per cent chance that freezing temperatures will occur *after* April 15.

Unfortunately, some people mistakenly use the average date of the last frost as the "safe" date for planting frost-sensitive crops such as tomatoes and corn. You can plant around that date if you are willing to accept a 50 per cent risk of freeze damage, but most growers are not quite that daring. By waiting approximately 10 days to two weeks beyond the average date of the last spring freeze, the risk of frost damage is reduced to about 25 per cent or less. About one month after the average date of the last 32° temperature, there is less than a 10 per cent probability of a freeze occurring.

FIRST EXPECTED FROST DATE IN THE FALL

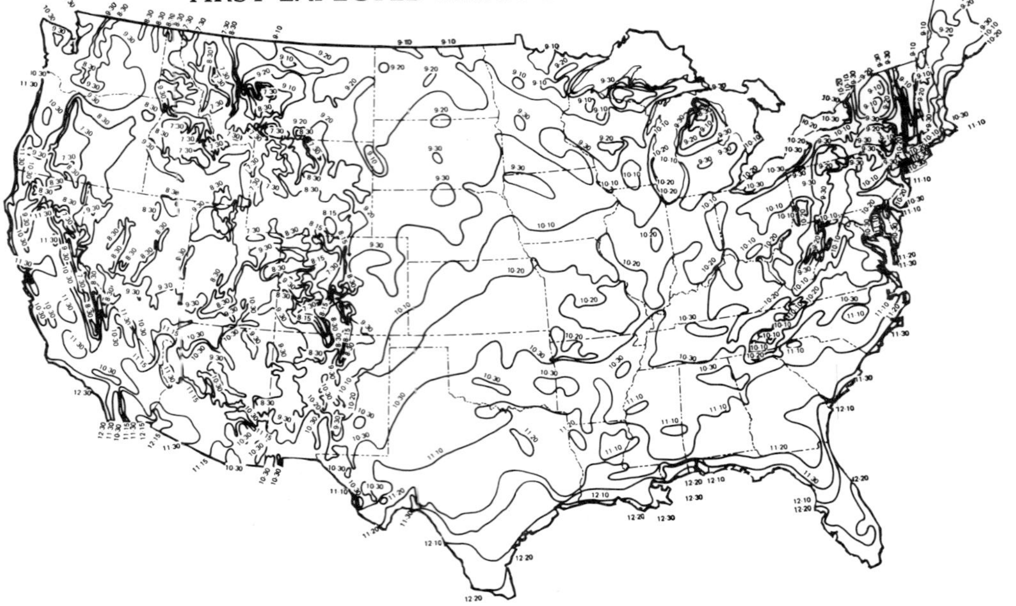

*This map gives a general indication of the month and date of the first frost or 32° temperature in the fall. Caution should be used in interpolating this map. Elevations or terrain features may cause frosts to occur earlier or later than indicated.*

In the fall, approximately 10 days to two weeks *before* the average date of the first 32° or lower reading the risk of frost in most areas is about 25 per cent, while there is less than a 10 per cent probability that frost will occur a month before the average date.

The average dates of the last spring and first fall frosts are shown on maps of the United States. These familiar maps should be used only as a general guide. On maps of this scale it is impossible to show the local variations, which often can be considerable (see Chapter 9). Growers who have lived in an area for some time become aware of local differences in minimum temperatures and the occurrence of "frost pockets." In most states, publications showing more detailed freeze data are available through Agricultural Experiment Stations or county agents. These data can be quite useful to farmers and gardeners. Even maps of this scale cannot account for all local differences, but they are much more useful than the larger national maps. An example of the type of map that can be made from statewide data is shown for Pennsylvania.

Some states have calculated the dates of 75, 25, and/or 10 per cent freeze probabilities in addition to the average (50 per cent) dates. Such information is extremely valuable because it helps growers select planting dates after which the risk of frost damage is at an acceptable, low level.

Perhaps a word or two about frost probabilities is in order. The average date of the last spring freeze, as mentioned earlier, represents the 50 per cent probability date. In other words, over a period of years, based on previous records, a grower may reasonably expect that about one year in two will have a frost after that date. This does not mean "every other year"—it does mean that over a ten-year period, for example, it is likely that about five of those years will have freezing temperatures after the average date. You might go four or five years in a row with no late freeze, and then get two or three consecutive years with a 32° or lower

AVERAGE DATE OF LAST 32°F. IN SPRING FOR PENNSYLVANIA

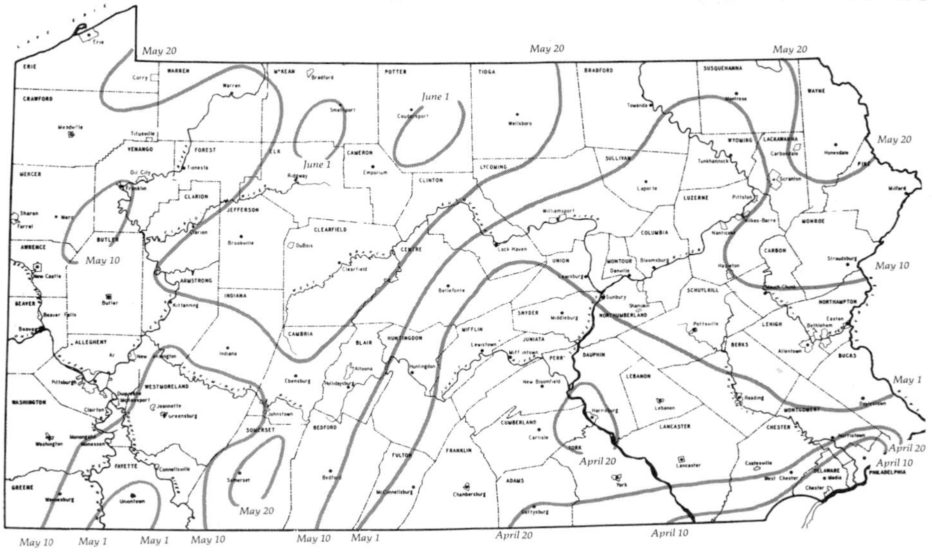


*This kind of map may be prepared for most states, using data from Agricultural Experiment Stations or other sources. But, even maps of this scale are only general guides; they cannot account for all local factors that influence temperatures, such as terrain and elevation.*

temperature after the average last-freeze date.

Likewise, the date with a 25 per cent probability of frost in spring indicates the date *after* which you can expect frost about one year in four—not necessarily every fourth year, but perhaps five years out of 20. The 10 percent probability date represents the date beyond which your risk of frost damage is only 10 per cent, indicating that it is likely only about one year in every ten. In the fall, the 25 per cent freeze probability date indicates the date *before* which there is likely to be a temperature of 32° or lower about one year in four. For example, if the 25 per cent freeze probability date for your area is October 1st, a 32° or lower temperature is likely to occur before the first of October in about five years out of 20. Similarly, in perhaps two years out of 20 a freeze will occur before the 10 per cent probability date.

While such probabilities cannot be used to forecast the date of the last spring or the first fall freeze in any given year, they do define the risk involved. Knowing this risk, you can then plan your planting dates and other activities accordingly.

## Defining Growing Seasons

Using the dates of various freeze probabilities also permits you to better define your true growing season. Normally, the average growing season is determined by the average dates of the last spring and the first fall frosts, and for many purposes this is probably satisfactory. However, if you are not willing to accept a 50 per cent risk of frost injury, then your growing season is shorter—especially for warm-season crops. For example, if you prefer to accept only a 25 per cent risk of frost damage in the spring and fall for certain sensitive crops, then your effective growing season for those crops becomes the length of time between the 25 per cent freeze probability dates, a period which could be as much as three to four weeks shorter than the conventional growing season. By obtaining freeze data for your own locality you can define your growing season much more accurately. If you have your own home weather station (Chapter 1), you can determine freeze dates and see how they differ from the

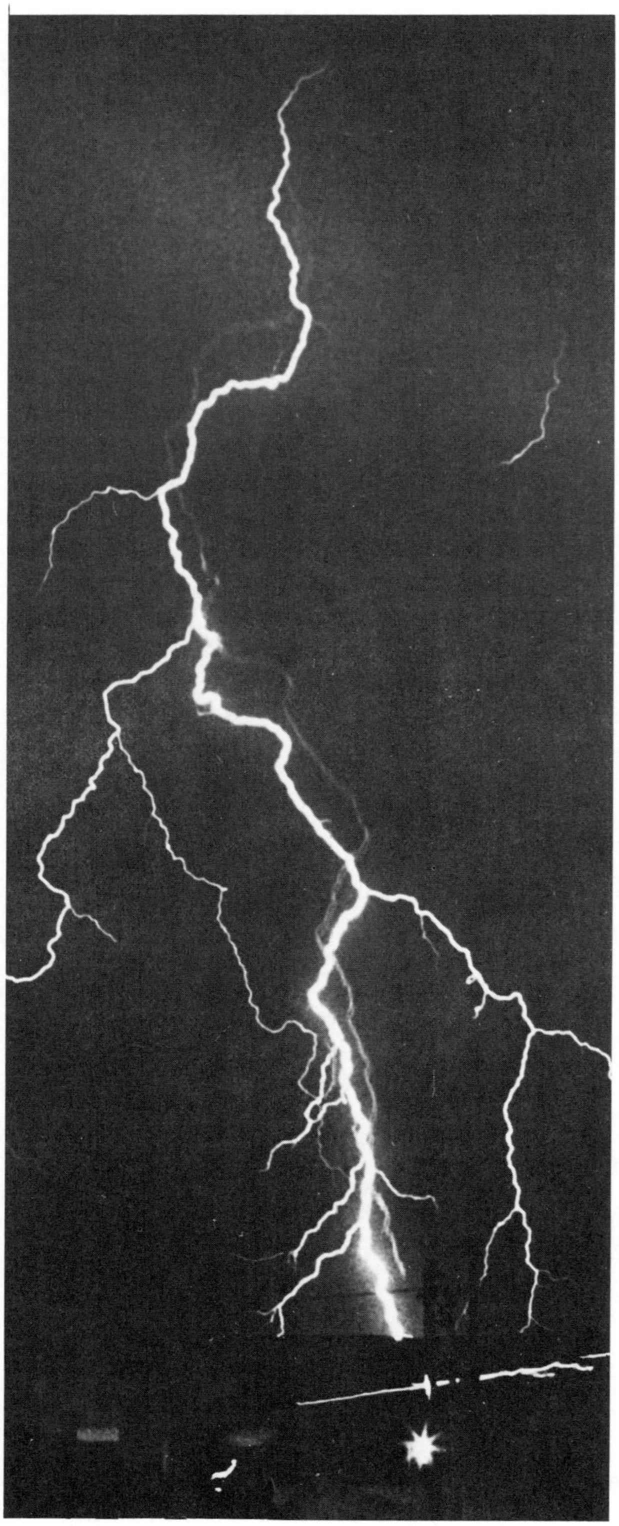

nearest official data. They might vary by as much as a week or more, especially in hilly or mountainous terrain.

Once you have defined your effective growing season accurately, you can make an intelligent selection of crop varieties for your location. For example, if you live in a cool climate with an effective growing season of 110 days, in planning the variety of winter squash to grow, you might select a Butternut or Acorn, which matures in about 75 to 85 days, rather than one of the Hubbards, which require 110 to 120 days.

Although freeze probability data based on temperatures of 32° or lower are most common, in many states information is also available on probability dates of other critical temperatures, such as 30°, 28°, and 26°. Some plants, such as orchard trees, are not affected significantly by 32° readings, depending on the stage of development, but have a critical temperature of perhaps 28° or 30°. Thus, probability dates for these lower temperatures are extremely important. Check with your county agent or your state agriculture department for published information for your area.

## RAINFALL STATISTICS

While freeze data are extremely important in agriculture, there are other significant weather parameters. Obviously, rainfall is critical. And the probability of a certain amount of rainfall occurring over a particular length of time can be calculated from climatological records. Such precipitation probabilities are especially important in determining the design and economic feasibility of irrigation systems. For example, if you know a particular crop requires at least 15 inches of moisture during its growing season for a satisfactory yield, an analysis of growing

NORMAL ANNUAL TOTAL PRECIPITATION (inches)

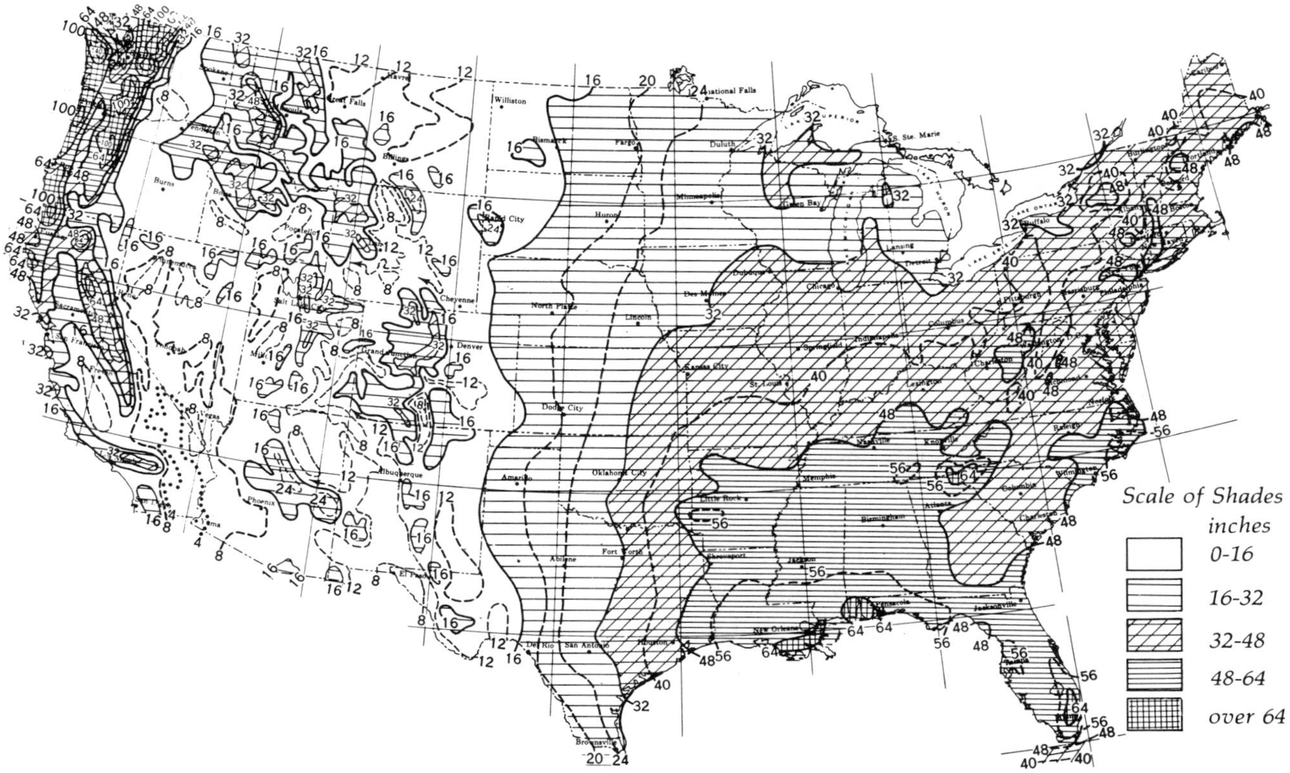


*This map, showing normal annual precipitation, may be helpful to gardeners and farmers trying to anticipate irrigation or watering needed for their vegetable crops.*

season rainfall can be made for the location in question. If the data show that at least 15 inches of precipitation can be expected in about eight years out of ten, there may be little economic advantage to installing irrigation equipment. If, on the other hand, there is a likelihood that only four or five years out of ten will receive the required 15 inches or more of moisture, then the installation of some type of irrigation system should be considered seriously. The final decision, of course, will depend on an analysis of costs and benefits.

Even knowing just average rainfall for your area, perhaps along with past extreme wet and dry conditions, can be a useful guide in selecting types and varieties of crops and even in selecting planting dates to take advantage of periods likely to have favorable moisture conditions. In some areas, data are available on the probability of various amounts of rainfall occurring each week or month of the growing season.

Another rainfall statistic that has been calculated for certain areas of the country is the probability of a selected number of consecutive wet or dry days. This information is valuable to hay growers and others concerned with drying conditions. It also helps determine when soil preparation can begin in spring and when to schedule fall harvest operations.

# GROWING DEGREE DAY VALUES AND TEMPERATURE DATA

Climatological data on growing degree days are especially useful to farmers selecting dent (field) corn varieties and will prove helpful to any grower, large or small, who determines the growing degree day requirements for his particular crops, as outlined in Chapter 6. Combining climatological data on growing degree days with freeze date probabilities can more sharply define the growing season length and the seasonal growing degree days actually available for use by a given crop.

For example, suppose a corn grower decides he will accept a 25 per cent frost risk in spring and fall. He determines that for his area the 25 per cent freeze probability date in spring is April 30 and the 25 per cent freeze risk date in fall is October 10. His effective growing season thus runs from April 30 to October 10, or 163 days. He then finds that the average growing degree day accumulation from April 30 to October 10 is about 2900 GDD. Thus, in selecting a seed corn variety, labelled in growing degree days, he should choose one that requires no more than 2900 GDD to reach maturity. The use of such a variety will not guarantee a perfect, undamaged crop, but will improve the chances of success.

For gardeners who do not have local growing degree day data, or have not determined the GDD requirements of particular crops, the same basic technique can still be used in selecting suitable crop varieties based on the calendar-day system of maturity rating. As an example, suppose you want to grow pumpkins and wish to reduce your risk of frost damage to 10 per cent or less in spring and 25 per cent in fall. If the 10 per cent freeze probability date in spring is May 25 for your area and the 25 per cent freeze risk date in fall is September 10, your effective growing season is about 108 days. Thus, you should select only varieties of pumpkin that carry a maturity rating of less than 108 days. And be sure to allow yourself a "safety factor" of at least a few days, in case the season turns out to be cooler than normal.

While growing degree day data are more specific, average monthly temperatures indicate the approximate rate at which particular crops are likely to mature. Where definite temperature limits have been established for successful growth of a crop variety, an analysis of local climatological data can help determine the feasibility of growing that crop in a given area. For example, if the crop requires overnight temperatures between 55° and 65° F. to set fruit, the probability of having a sufficient number of nights with temperatures in that range during the critical fruit-setting period can be determined. Likewise, if a particular crop is sensitive to temperatures above 90°, an analysis of climatological data can easily provide the average number of days with maximum temperatures 90° or higher during the growing season. The probability that no 90° plus days will occur, or the probability of a particular number of days with maxima over 90° can also be determined.

# OTHER CLIMATOLOGICAL DATA

Other weather probabilities are also important in many ways to agriculture, although the data on freeze dates, rainfall, growing degree days, and temperature are probably the most widely used. Knowing the probability of hail damage is critical to growers in selecting crop types, to the horticulturist in determining the risk of damage to his greenhouse, and to insurance companies which write policies to cover hail storm losses. Information on wind direction and speeds is needed when constructing windbreaks or animal shelters. Data on average and extreme relative humidities as well as air temperature are needed to make accurate decisions on air conditioning for animal housing.

Information on all these weather parameters, and others, can be used by growers to make rational decisions a season or more in advance. While the farmer or home gardener (or the general public, for that matter) can not obtain accurate weather forecasts weeks or months in advance, they can use climatological data as an indication of the risks associated with various choices available to them.

## DATA AVAILABILITY

Growers with home weather stations will be able to calculate average temperatures, rainfall totals, freeze dates and other statistics for their area once they have accumulated several years of weather data. An example showing how to calculate average temperature, rainfall totals, and freeze dates is shown in Appendix H. Determining the number of days with temperatures above or below some critical value, the extremes of temperature and rainfall for each month of the growing season, the longest period without measureable rainfall, and other similar statistics from your own daily weather record is quite easy. Developing more involved data, however, such as probability values, requires the services of those experienced in statistical analysis and interpretation. Most growers must therefore rely on published values of such data for their area or hire a professional climatologist to analyse available records.

Climatological information is available from a variety of sources. In many states, weather data can be obtained from county agents, agricultural experiment stations, or agricultural colleges and universities. Farm and garden publications may publish such data occasionally. Local National Weather Service offices will usually have data for their immediate area and sometimes for other regions of their state. The national *Weekly Weather and Crop Bulletin* publishes selected national data, while many states publish their own weekly crop and weather reports, at least during the growing season. Most material from these state and local sources is free or available for a nominal charge.

Should you desire data not available from these sources, you can contact the National Climatic Center in Asheville, N.C. This facility is the repository for all climatological data collected in the United States, and some foreign countries as well. Describe the type of data you require and the area you are interested in. They will let you know if that type of information is available for a nearby station. This service can be expensive, so with your inquiry request an estimate of the total cost for providing the data. In some areas, private climatologists are available. While the basic data they supply would generally be the same as that obtained from the National Climatic Center, they may be able to provide specific data analysis and interpretation, particularly for areas with which they are familiar.

A more detailed discussion and listing of sources for climatological data is given in Appendix C.

# 9
# The Microclimate

All of us, whether at home, work or play are affected by, and are a part of, the local microclimate. Microclimate refers to very small-scale, localized environmental conditions which are affected by local topography, soil conditions, buildings, vegetation, etc. Technically, even the indoor climate of buildings may be considered a separate microclimate, but here we are interested only in natural outdoor conditions. Alert gardeners and farmers should pay close attention to the microclimate before selecting a planting site.

There are three scales of climatic activity. The macroscale climate (or ***macroclimate***) refers to climatic conditions over large areas, such as the central United States, eastern North America, Western Europe, and the North Atlantic. When people speak of the climate of an area they are usually referring to the large-scale, or macroclimate.

The term ***mesoclimate*** is used to describe more localized weather conditions, such as the climate in and around a large metropolitan city, or around a large lake. The heavy "lake-effect" snow squalls common along the lee shores of the Great Lakes in late fall and winter, and which occasionally paralyze cities such as Erie, Pennsylvania and Buffalo, New York, are examples of mesoscale events. The "heat island" phenomena observed in large urban centers is another example of a mesoclimatic feature.

Finally, the ***microclimate*** refers to the unique climatic conditions which may exist on an individual farm, around a small lake or pond, or in a backyard garden. Frost pockets are a feature of the microclimate.

Why should you, a home gardener or farmer, be concerned with your local microclimate? Mainly because variations found in local climatic conditions can mean the difference between success and failure of a crop, particularly during critical low-temperature periods. Let's take a brief look at how the local microclimate affects specific weather parameters.

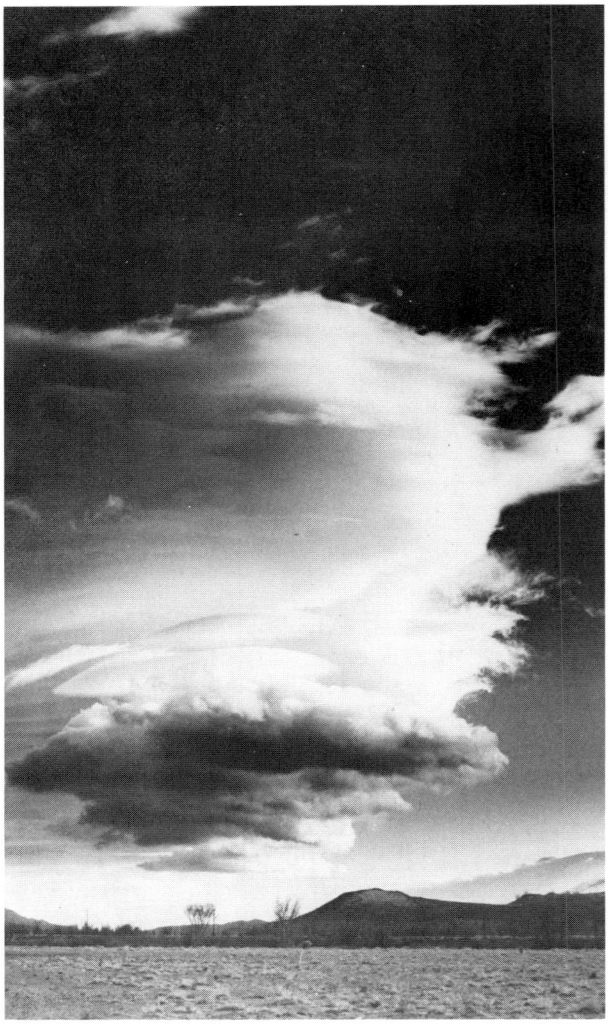

*An unusual cloud formation over rural California, indicating severe upper air turbulence.*

# TEMPERATURE

In general, the weather element most commonly affected by local conditions is the temperature—particularly the overnight minimum temperature. During the day, as the sun's energy heats the surface (either natural or artificial—soil or asphalt) the air next to the surface also becomes heated and, as it warms, it becomes lighter and rises. Cooler air at higher levels descends to replace this warm air, creating what are known as convection currents. Thus, daytime heating usually mixes the air near the earth's surface well, and horizontal temperature differences tend to be smoothed out. The greatest temperature differences exist at or very near the surface, because the type of surface affects air temperatures immediately above it. Temperatures over bare soil, for example, will be higher than those over a sod cover. Air over an asphalt parking lot is noticeably warmer than air over a grass-covered field.

At night, however, there is no surface heating to mix the air and, unless a good breeze is blowing, the air has a tendency to become "layered." As various surfaces, such as soil, plant, and concrete, lose heat by radiation, the air next to them also cools. Since cold air is heavier than warm air, the coldest air remains at or very near the surface. Thus, while temperatures generally decrease with height during the day, at night the reverse is often true, with the coldest air found next to the ground. This condition is called an ***inversion.*** Depending on the overall weather conditions, the inversion may be very shallow or may extend upwards for hundreds or thousands of feet.

Above the top of the inversion, temperatures begin to decrease with height in the more normal manner. A plotted curve showing how the temperature changes with height is sometimes referred to as a ***temperature profile.*** The figure at right illustrates the normal temperature profiles for both daytime and nighttime conditions, with generally clear skies and light or calm winds.

The overnight sky and wind conditions are critical. If a good breeze is blowing, the air remains fairly well mixed and the milder air aloft mixes with the colder air near the surface to modify the temperatures in the lower layers. Likewise, under cloudy skies the radiation from soil, plant, and other surfaces is reflected back towards the earth by the clouds and very little cooling occurs. Thus, with strong winds and/or a substantial cloud cover, nighttime temperatures are usually rather uniform over large areas and microclimate differences are small. It is the clear, calm nights during which local topography and surface conditions can combine to create substantial differences in minimum temperatures over short distances. Such nights are often referred to as "good radiation nights."

As we saw earlier, cold air is heavier than warm air and will settle near the surface. If there is a slope to the surface the coldest air will drain down the slope just as water flows downhill. Thus, on good radiation nights, the coldest temperatures almost always occur in valleys and other low-lying areas. It is not uncommon to have minimum temperatures in valley areas that are five to ten degrees Fahrenheit lower than those on nearby hillsides.

TEMPERATURE CHANGES
ABOVE THE SURFACE
WITH CLEAR SKIES AND LIGHT WINDS

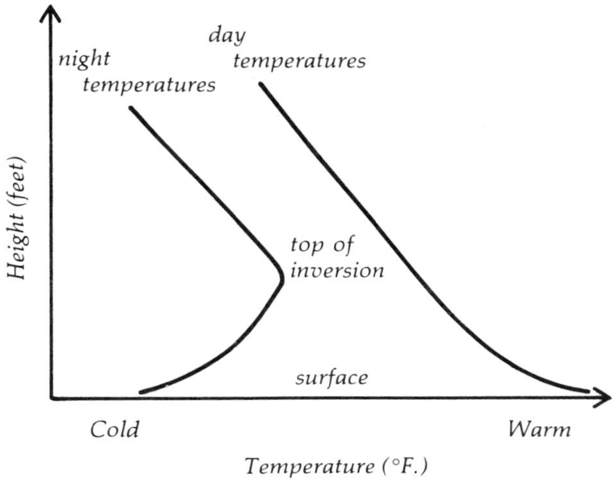


*While temperatures gradually decrease with height during the day (right line), at night the reverse is true, with the coldest air found next to the ground (left line). This condition is called an* inversion. *Above the top of the inversion, temperatures begin to decrease with height.*

## Vertical Temperature Differences

Not only do large temperature differences sometimes occur within short horizontal distances, but there can also be a sharp vertical difference in temperature within the first few feet above the surface. Temperature differences of five to ten degrees Fahrenheit between the surface and a standard weather shelter are not unusual on clear, calm nights. That is the main reason why a standardized shelter height of about 5½ feet has been chosen for official climatological stations, as was noted in Chapter 1.

Even slightly rolling land can produce important temperature differences, and for home gardeners this type of topography is common. Small dips or hollows in a generally flat landscape will collect cold air draining from the higher terrain and may have temperatures several degrees lower than most surrounding areas. I had an interesting opportunity to take part in a study which measured overnight minimum temperatures in an apple orchard in southern Ontario during several critical spring cold spells. Significant temperature differences were found within the orchard, as shown in Table 14. The differences in elevation between the highest and lowest points (knolls and basins in Table 14) in the orchard where the temperatures were measured were only several feet, but the variation in temperature was sufficient to produce noticeable differences in freeze damage to the apple buds. Note also that the grass surface within the orchard averaged eight to nine degrees colder than the air at the five-foot level.

### Table 14
### Minimum Temperatures (°F.) in an Apple Orchard

| | 5' Above Surface | | Grass Surface | |
|---|---|---|---|---|
| Date | Knoll | Basin | Knoll | Basin |
| April 27 | 23 | 18 | 16 | 11 |
| April 28 | 24 | 17 | 15 | 11 |
| May 2 | 26 | 23 | 19 | 15 |
| May 6 | 25 | 22 | 15 | 9 |
| May 7 | 22 | 18 | 13 | 9 |
| *Average* | 24 | 20 | 16 | 11 |

The alert gardener or orchardist should thus pay close attention to the local topography before deciding on a planting site. Of course other factors must also be taken into account—factors such as soil conditions, ease of accessibility and water availability. But, when picking a site, try to avoid obvious "frost pockets." A

EFFECT OF TERRAIN ON MINIMUM TEMPERATURES
ON CLEAR, CALM NIGHTS

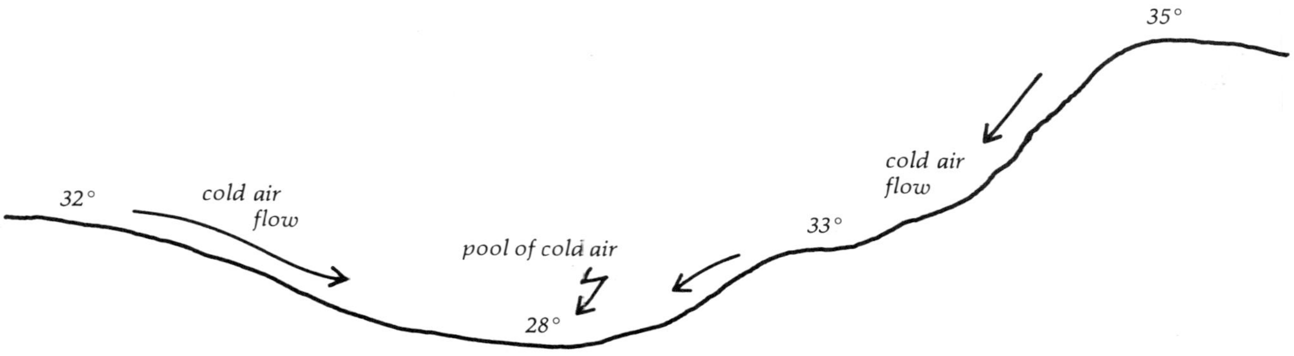


*Slightly rolling land can produce important temperature differences. Small dips or hollows will collect cold air draining from higher terrain. These pockets should not be used as gardening sites.*

few degrees difference in temperature on a cold night might make the difference between crop survival or crop loss. See Chapter 10 for a more detailed discussion of crop site selection.

Incidentally, the sharp differences in minimum temperatures on clear, calm nights are a good reason for having your own maximum/minimum thermometer and developing your own technique for estimating frost danger, as described in Chapter 5. General weather forecasts covering large areas just cannot take into account those local differences.

Bodies of water produce unique local climates which can be used to good advantage by growers. Because water gains and loses heat less rapidly than does the land surface, the area adjacent to a lake will have a more moderate climate—generally milder in winter (assuming the water is unfrozen) and cooler in summer. (Once a lake becomes frozen, and perhaps even snow-covered, it behaves as any other frozen surface and loses its moderating influence.) The water moderates cold nighttime temperatures and helps prolong the growing season. Orchards and vineyards are quite common along the shores of Lakes Ontario and Erie, despite their northern latitude and the more severe climate found a short distance away from the water. In addition to extending the growing season, the cooler waters in spring retard development of the plants and thus protect against late season freezes. By the time blossoming finally occurs, the chances of a damaging frost are quite small. The amount of protection provided by a body of water depends mainly on its size and depth, but even small ponds may moderate the temperatures a short distance "inland." Evaporation from a lake will also increase the amount of moisture in the air over the surrounding countryside.

# PRECIPITATION

Local differences in rainfall within a region can be quite large. For example, the June 1974 rainfall recorded at our garden site in Camp Hill, Pennsylvania was 5.54 inches, while the official National Weather Service measurement taken at an airport only 4.25 miles away was only 3.69 inches—almost two inches less. During one three-day period the official total was 0.93 inch, while our garden gage recorded 1.82 inches, or nearly twice as much. A recent study in the mountain and valley region of West Virginia found a difference of 35 inches in the average annual precipitation between two stations within a horizontal distance of 45 miles.[1]

Once again these examples illustrate the importance of having your own raingage to *determine actual precipitation at your garden site.* A decision to irrigate based on the nearest official rainfall measurements could cause you to either overwater your plot or be too late with the additional moisture.

A key factor affecting local differences in rainfall is topography. As air is forced upward over large hills and mountains it is cooled, moisture condenses out, and clouds are formed. If enough moisture is present and the air is forced high enough, precipitation will occur. The figure, *Terrain Effects on Precipitation,* shows this process. On the downwind side of a hill air descends, is warmed, and becomes drier. Thus the downwind (lee) side of a range of mountains will generally be dry. Good examples of this phenomena are found in the western United States, where the Coastal

*Terrain Effects on Precipitation*

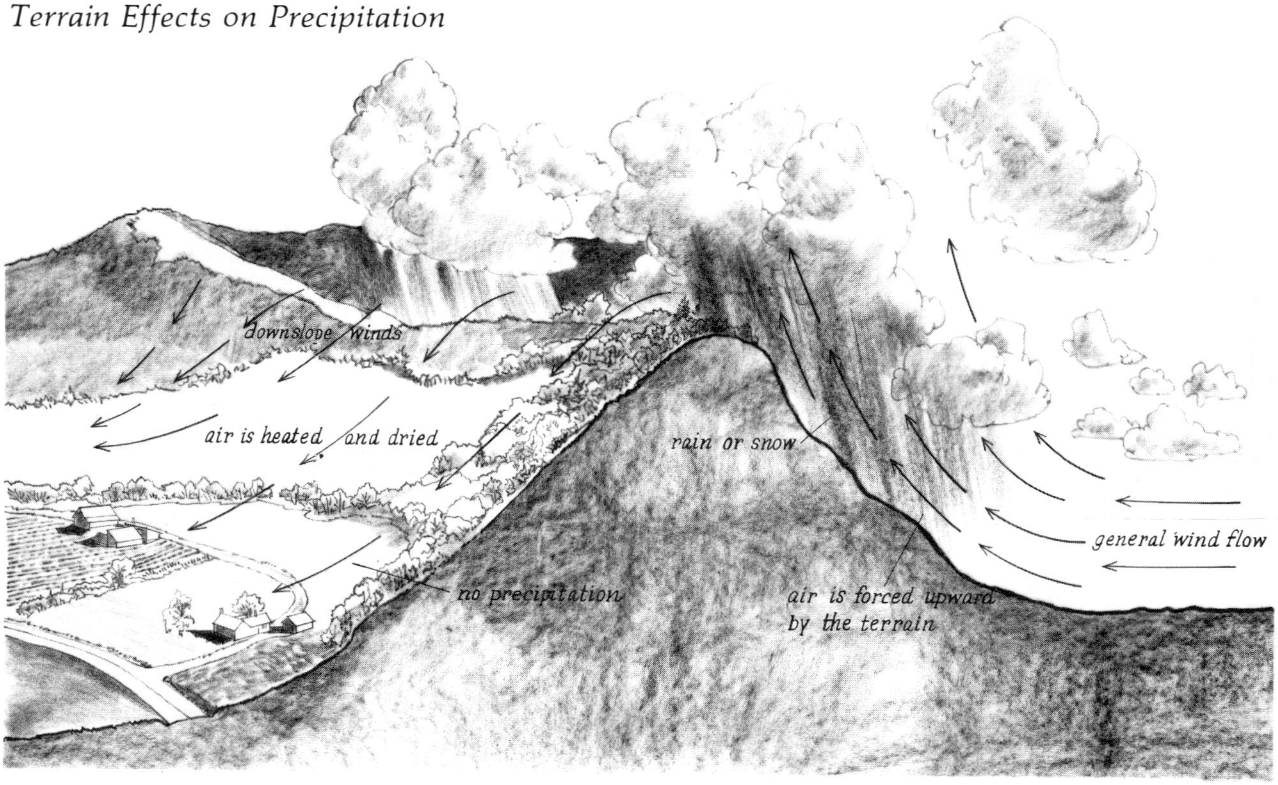


*As air is forced upward over large hills it cools, moisture condenses out and clouds form; if enough moisture is present and the air is forced high enough, precipitation occurs.*

*On the downwind side of the hill, air descends, warms and becomes drier.*

Range, the Cascades, and the Sierra Nevada produce alternating series of high rainfall regions and rain shadows. A similar, but less pronounced, precipitation pattern can be found in some of the mountainous areas of the eastern United States.

In regions where terrain differences are small, with only occasional low hills, or on completely flat land such as the Great Plains region, the effect of topography on rainfall is negligible. Nevertheless, large local differences in rainfall may still occur, especially with summertime showers and thunderstorms. The primary difference is that, whereas in mountainous terrain there are preferred areas which repeatedly receive either an excess or a deficiency of rainfall, over flat terrain the areas of showers are more random and often vary from one rainfall period to the next. The reasons local showers develop over a particular area in flat terrain are still not very well understood, although differences in the rate of heating of the ground surface may be an important factor. Much research remains to be done in this field.

# WIND

In general, on land that is gently rolling, the topography will not have much effect on the local winds—at least not enough to be of concern for gardeners or farmers who are interested in wind conditions for spraying or dusting operations. In more rugged country, however, with sharp hills and valleys, the landscape will affect the local wind flow, at least under certain conditions.

Usually, if the large-scale wind flow over an area is strong, it will override any effect of local heating or cooling on the development of small-scale wind patterns. The winds will tend to "funnel" through valley areas and be deflected around hills. If the general air movement is light, however, local breezes may develop that reflect differences in topography and/or heating or cooling.

During the middle part of the day, particularly with bright sunshine, heating of the ground surface causes the air to rise and produces a flow of air up the sides of slopes and up through valleys. On generally clear, calm nights, as the air next to the surface cools, a downslope and down-valley wind will develop. The speed of these winds will depend on the degree of the slope, as well as on the amount of heating or cooling of the ground surface relative to nearby areas.

Another type of local wind is the sea or lake breeze. This is generally most pronounced during midday in the warmer months of the year. As the land surface warms much more rapidly than the water, a circulation of air develops as shown in the figure, *Lake Or Sea Breeze.* This circulation produces cool and usually quite welcome breezes over the land areas next to the water. Generally, the greater the difference in temperature between the land and the water, the stronger the winds. A reverse circulation of air can occur at night as the land cools more than the water, although it is usually less pronounced. Again, these local wind conditions can be overridden by a strong flow of air around large-scale weather systems.

LAKE OR SEA BREEZE

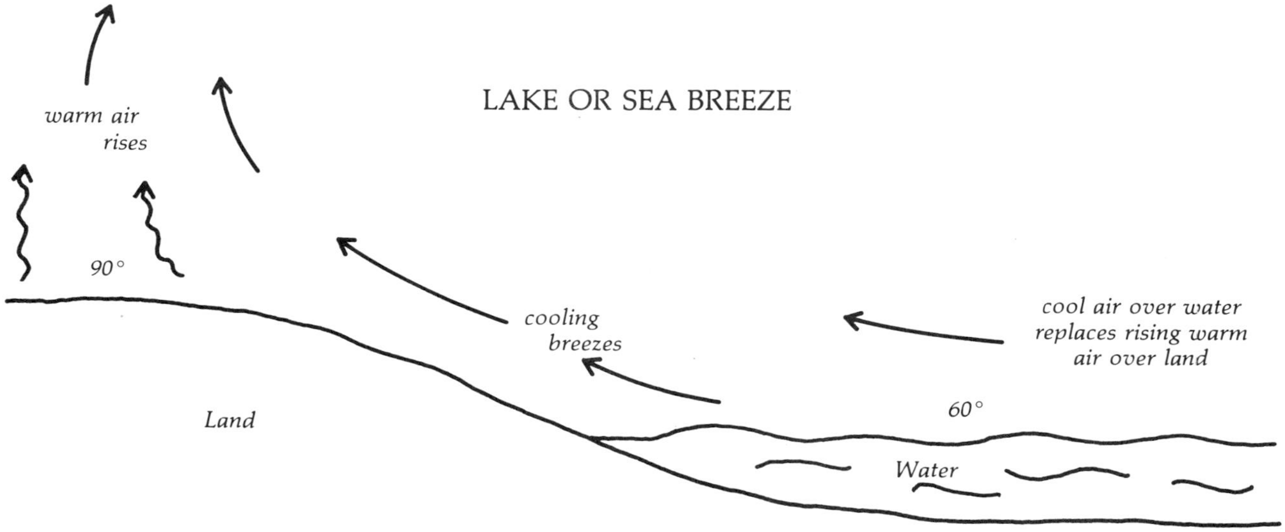


*Cool lake or sea breezes sometimes develop at midday when land surfaces warm more rapidly than the water, producing a favorable, pleasant air circulation.*

# 10
# Weather and Site Selection

In selecting a home and/or a garden site, most of us are limited: the site must be on property we own or rent, or on property we can afford to purchase. Many factors must be considered when buying a piece of land, or when building a home or laying-out a garden plan on land already owned. Despite their importance, weather and climate are factors that are not often considered. This chapter takes a brief look at how you can use your basic knowledge of weather and climate to locate your garden or orchard to maximize production and minimize the risk of frost and other weather hazards. How to better plan your home for the most efficient energy use is also examined.

# GARDEN AND ORCHARD PLANNING

Often the selection of a garden site is dictated more by aesthetic than by environmental considerations. Some people prefer to put their garden behind the garage or in some out-of-the-way area where it won't be too noticeable. Others will plan their overall landscaping and then put in a garden if, and where, space permits. Fruit trees are often located where they will provide the best "show" during their spring blossoming period, or where they will act as good shade trees later in the spring and summer.

Unfortunately, these reasons for locating a garden or group of fruit trees at a particular site may not provide the best "fit" with the local microclimate. By using some of the information presented in the previous chapter on local "mini-climates," you can plan your garden or orchard site to take maximum advantage of your environmental conditions.

## Garden Sites

The ideal location for a garden plot would be on a nearly level piece of land, perhaps at the top of a very small rise, with the ground gently sloping away from the site. The area should be open, with little or no shading. Such a location would permit early warming of the soil in spring, provide minimum frost risk, and allow good drainage.

Unfortunately, the ideal garden site may not be available, but you can at least use the land you do have to its best advantage. There are

*This garden site is almost ideal: It has a gentle, southerly slope; it's protected from northerly winds; and it receives nearly full daytime sunlight.*

*Another good planting site is this one in West Townsend, Vermont, with a gentle slope, favorable sunlight and wind barriers on several sides.*

several basic rules to follow in selecting a suitable garden site.

First, avoid the lowest point on hilly property, as well as the occasional hollow on otherwise flat land. Since cold air drains downhill just like water (see discussion and figure in Chapter 9) these low spots become frost pockets on clear, calm nights in spring and fall when temperatures drop toward the critical freezing mark. Try to picture water flowing over your property and visualize where it would form pools. (During heavy rainfalls you may see such pools develop.) Those pool areas collect cold air, as well as water, and should be avoided if possible when planning your garden. Or at least save such spots for hardy or mid-summer crops when frost is not a problem.

Secondly, if your garden plot cannot be on level land, a gentle southern slope would be your next best bet. Southern slopes will receive the best sunlight and the most heat; soil will heat quickly in the spring. And more heat in the soil during the day will mean more heat will be released at night, keeping surface temperatures higher near your plants. Crops located on the upper parts of slopes will be less susceptible to frost damage than those at the bottom.

A plot on the higher part of your land will also provide better drainage (soil conditions being equal) and is less likely to puddle should there be excessive rainfall. If the slope is steep, garden rows should be parallel (horizontal) to the slope and perhaps even some form of terracing would be advisable to prevent erosion. Keep in mind that soil that drains exceptionally well is more subject to the effects of drought, so you will need to be extra alert to the water needs of your crop during dry periods. Mixing large amounts of organic matter into the soil will improve its water-holding capacity significantly while still permitting excess water to drain away following heavy rains.

*A well-designed "diversion dam" built of hedges, walls, or fences helps divert cold air away from the garden.*

If your garden must be near the bottom of a slope, erect some type of fence above your plot—either artificial or natural. A dense hedge would be ideal in most situations. This fence will act as a dam to help hold back cold air draining from the upper part of the slope and provide at least some garden protection. By shaping the fence properly, the cold air can be drained away from the garden area.

Also, a windbreak of trees or a hedge on the prevailing wind side of your garden (usually the southwest, west, or northwest side—but check the data for your particular locality) will reduce water loss from evapotranspiration* and also help minimize wind damage during severe storms. The soil moisture saved because of the reduced evapotranspiration could be critical during drought periods.

*Evapotranspiration is the total water lost by evaporation from soil and transpiration by plants.

## Orchard Sites

Orchard or berry-patch site selection is based on somewhat different criteria than those used for selection of a vegetable-garden plot. Once again, the lowest parts of the land, with possible frost pockets, should be avoided. Fruit trees will generally be less prone to frost damage if they are grown on the upper part of a slope. However, *northern* slopes are preferable to southern exposures. Plants on the southern slope will warm faster in the spring and begin development earlier than those on the north side. If a late spring freeze occurs, trees most advanced in their development will usually suffer the greatest damage, since tightly closed fruit buds can generally withstand lower temperatures better than can open buds. In this case, then, the less rapid warming found on northern slopes in the spring is a help, rather than a hindrance.

Other factors, such as soil type, drainage, and accessibility to the plot, will of course enter into your selection of a garden or orchard location. But do consider your local microclimate. A little forethought could extend your growing season significantly, increase your yields, and reduce the risk of weather related losses.

## SELECTING A HOME SITE

As we have already noted, there are many factors which enter into either the selection of a site on which to build a house, or the purchase of a piece of property on which a dwelling is already located. One of these factors is (or should be) the local climate and weather conditions. By building your house in harmony with the environment you can significantly improve the energy efficiency of your heating and cooling systems and generally have a more enjoyable and natural landscape. And by facing your greenhouse or plant room to take advantage of available sunlight, you can increase the amount of indoor heat for beginning vegetable plants in the spring.

If you have a choice in the orientation of your house, consider placing your living room or family room, kitchen, and other well-used areas on the south and west sides. This arrangement will provide the maximum amount of light during the day—especially on dark winter days when the sun is low and clouds are prevalent in many areas of the country. Other rooms, such as the bedrooms, which receive little use during the day can be on the north and east sides. With this arrangement the need for artificial light during the day will be reduced and electric bills will be at least a little lower.

If possible, avoid large picture windows facing northwest, north or northeast. Little or no sun enters a home from these directions, but cold winter winds from these directions are common. If your home already has a northerly window, you can reduce heat loss by planting a protective windbreak of evergreens.

An alternative, of course, would be to do some major remodeling and eliminate any large windows which face away from the low winter sun. Such an extensive renovation would involve quite an initial expense, but eventually would pay for itself in fuel savings. Even south-facing windows lose considerable amounts of heat on cloudy days and at night. The use of double- or even triple-pane glass reduces such heat loss, as does the hanging of heavy drapes inside the windows. The drapes can be opened on bright sunny days and closed at other times. Large windows facing south or southwest, while admitting welcome sunshine during the winter months, add an extra load to air conditioning systems during the warmer part of the year. Again, the use of double- or triple-pane glass and heavy drapes can reduce this heat load. In the extreme south and southwestern parts of the United States, even the winter sun can produce excessive warmth at times.

One way to benefit from winter sun while eliminating unwanted heat during the summer is through the proper use of awnings or roof overhangs. Roofs or awnings on the south and west sides of the house can be positioned to allow the low winter sun to shine through the windows, providing welcome heat and light, while shading those same windows from the hot and higher summer sun. The exact amount of overhang and the angle at which it should be placed depend on your latitude, as well as the size and position of your windows. Each house is a special case. While the handy do-it-yourselfer can probably do a respectable job, the best results will be obtained if you hire a local architect—especially one who is well versed in "environmental engineering."

In solar heating and cooling, the orientation of a building becomes even more critical. Some of the newer books on solar and other alternate energy sources provide detailed plans for designing buildings to account for local weather and climatic conditions (see Appendix D). Many of the suggestions in these books can be used in the modification of existing dwellings as well as in the construction of new homes.

OVERHANGS FOR SUNLIGHT CONTROL

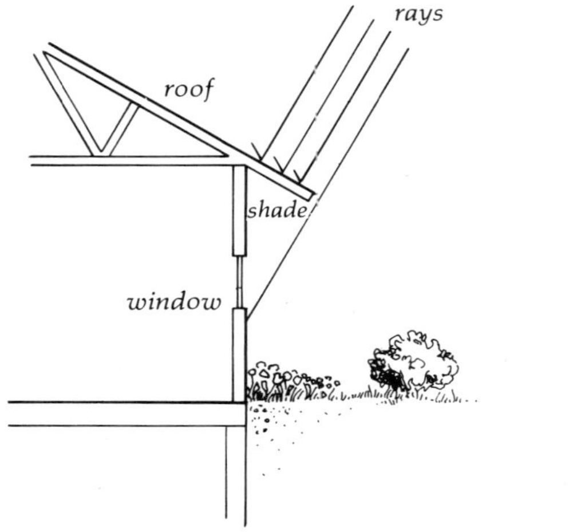


*Higher summer sun is bounced off the shade.*

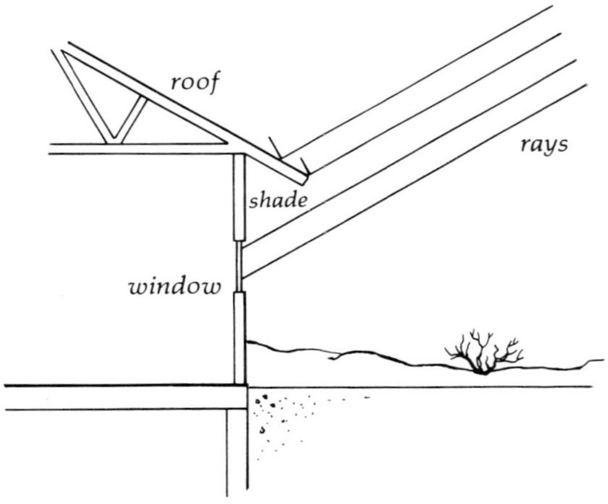


*Lower winter sunlight reaches the window.*

## LANDSCAPING

Landscaping can also be used effectively to ameliorate the effects of weather. A windbreak of evergreen trees on the north side of the house (or, even better, from the northwest around through the northeast) will significantly reduce the flow of cold winter winds against the outside walls and thus reduce heat loss. A recent study at Penn State University, for example, suggests that a properly placed shelterbelt could reduce the amount of fuel needed for heating by as much as 15 to 23 per cent—a significant savings at today's fuel prices. Another study in South Dakota, using identical houses—one completely exposed to winter winds, the other protected by a combination of trees and shrubs—found that the protected house used about 40 per cent less fuel than did its unprotected "twin." In milder climates, fuel savings of at least 10 per cent or more have been reported with the use of some type of windbreak.

The more dense the planting the greater the protection from cold arctic blasts. While even one row of evergreen trees or shrubs will provide a noticeable reduction in the wind flow,

*Dense planting will protect a farm or home from cold air blasts. This cedar windbreak in Tamora, Nebraska was planted in 1880.*

the use of two or more rows of plants produces significantly greater protection.

The distance of the windbreak from the house is critical. While some reduction in the wind flow can be found out as far as 25 to 35 times the height of the trees, the greatest reduction occurs at a distance of approximately four times the height of the windbreak. The actual distance depends on the plant type and planting density. Thus, for example, a shelterbelt of evergreens that can be expected to reach about 30 feet in height at maturity, should be planted about 100 to 125 feet upwind from the house to provide the greatest wind reduction.

Shading during the summer likewise has a significant effect on air conditioning requirements. South-facing rooms with windows shaded by large trees may be as much as 10° to 20° cooler than similar rooms with unshaded windows. Various tests have found that unshaded houses may require as much as 35 to 65 per cent more electricity for air conditioning purposes than do well-shaded homes. While evergreen trees on the sunny side of the house would provide cooling summer shade effectively, they would also prevent the welcome winter sun from warming the house. For this reason, deciduous trees are recommended on the west and south sides of the house. They will provide cooling shade in the summer but, bare of leaves, will allow the low winter sunlight to easily penetrate through to the windows and outside walls of the house.

So if you are planning to landscape a new home or are considering re-landscaping an older house, take into account the energy aspects of your overall plan. Investing in the services of a professional landscape architect might prove worthwhile in the long run if you have any doubt about the best and most energy efficient plan for your house.

# 11
# Climatic Change—Fact or Fiction?

During the first half of this century, worldwide temperatures appeared to be slowly rising and concern was increasing because man's industrial activities were adding more carbon dioxide ($CO_2$) into the air. There was quite a bit of speculation about the earth's climate warming excessively, resulting in all kinds of dire consequences. Sunday newspaper supplements and popular literature carried forebodings of increased warmth which would melt the earth's glaciers and icecaps, raise sea level by hundreds of feet, and flood populous coastal areas.

Then, in the 1950s, temperatures in the northern hemisphere halted their upward trend and, instead, began a slight decline. Suddenly, the "melt the glaciers and flood the coasts" theory became unpopular. In the 1970s, with the slow cooling trend persisting over many areas, and with unusual and unfavorable weather conditions suddenly assuming great importance because of their effects on the world's food and fuel supplies, a new popular theory emerged—the "another ice age is just around the corner" theory. Popular and scientific literature is filled with various predictions and explanations about the impending "deep freeze." A lengthy and well-done television program on climatic change left viewers with the strong impression that the only questions remaining were "When will it occur?", and "What will cause it?"—not, "Will it happen?"

So we have a choice. We can still fall back, for example, on the "earth is getting warmer" theory (which includes some good arguments) and say that the current cooling is only a minor pause in a long-term warming trend. Or, we can go along with the climatic cooling theory (which also has some valid points) and argue that the previous warming trend was only a brief respite from a long-term cooling of the earth.

Confusing? Probably so. What the argument really points up is basically that the earth's climate is never static but changes constantly. The changes may sometimes be long-term and virtually imperceptible, but they are there nevertheless. Weather and climate are very seldom "normal." Furthermore, weather trends in one region may be balanced by opposite trends in another. For example, a slow cooling trend has occurred throughout most of the country during the past two decades, while there's been a slight warming trend in parts of the Southwest.

Many cycles occur in climatic conditions. Some of them, like the glacial periods, last for tens of thousands of years. Others are short-term fluctuations, such as periods of drought or temperature change lasting perhaps only a decade or less. A good example of a short-term climatic change was noted in the northeastern United States, where record, crop-killing drought seared the land during the 1960s. During the 1970s, the same region has had monthly and annual rainfall records broken at many locations, two major widespread floods, and frequent flash-flooding of small streams. Likewise, a severe two-year drought in California and other parts of the western United States was broken, at least temporarily, by record rainfall during the 1977-78 season.[1]

*The history of the earth's climate includes a series of glacial advances over Europe and North America. This is a relatively small "expanded foot" glacier with vertical terminal walls about 33 feet high.*

All of these cycles merge and produce what we commonly call "normal" climate. At present, knowledge of the complex working of the atmosphere, and the effects of outside influences such as solar energy and cosmic rays, is insufficient to reconstruct accurately the causes of past climatic changes, and therefore our ability to forecast future changes is virtually nil. Nevertheless, research is going on in this field and perhaps one day we will at least have a better understanding of how and why the climate fluctuates.

## CLIMATIC HISTORY

A complete, detailed history of the earth's climate, would fill a large book and would be of interest primarily to specialists working in the field of climatology or related disciplines. However, a brief summary of climatic trends during the past 10,000 years or so will help put modern climatic conditions in proper perspective, and hopefully will be of some general interest.

The series of glacial advances which we refer to as the Ice Age began one to two million years ago. There were at least four separate major advances and subsequent retreats of the ice sheets in Europe and North America, with numerous minor fluctuations. The last of these ice sheets retreated about ten thousand years ago, and since then we have been in a relatively warm, or ***interglacial*** period. Those who believe we are entering a new cold period note that current evidence indicates past interglacial periods have lasted approximately ten thousand years, and thus our current one may be near an end. Until we can understand the causes of such changes, however, we just do not know if our present favorable climate will follow similar patterns in the past or extend indefinitely into the future. It is believed, however, based on analyses of past climates, that temperatures have been as mild as they are now only about five to ten per cent of the time during the past one to two million years. And only a few periods in the past ten thousand years have been as warm as the early and mid-1900s. So we could say that we are living in a period of favorable weather conditions known as a climatic "optimum."

## Changing Climatic Conditions

During the past ten thousand years there have been numerous minor changes in climatic conditions—minor, at least, compared to changes between glacial and interglacial periods, but important from the standpoint of their effects on human activities. From about 8000 to 5600 B.C. both Europe and North America apparently were generally cool, with a trend from moist to dry conditions during the period. Some minor readvance of the ice sheets may have occurred in North America during the latter part of this time span.

The period from approximately 5600 to 2500 B.C. is known as the Atlantic Period and was the first so-called "Climatic Optimum" since the end of the previous glacial period. Generally warm, moist conditions prevailed, with particularly wet weather noted in the area around the current Sahara Desert. Evidence of lush vegetation and a wide variety of animals, including elephant, hippopotamus and rhinoceros, during this period, have been found in northern Africa. There is also evidence that during this period, in the Mid-East, previously nomadic groups of people began to domesticate animals and plants and develop a food-producing economy—the beginnings of what we today call agriculture.

Toward the end of this period of favorable climate the weather turned drier than in the early optimum period, although still remaining warm. These warm, dry conditions persisted until about 500 B.C., when much cooler and wetter weather developed. Expansion of European glaciers occurred from 500 B.C. until around 400 A.D. Then, at least in parts of the Northern Hemisphere, a gradual improvement was noted in weather conditions, resulting in a second Climatic Optimum around 1000 to 1200

*Blue Glacier, Olympic Park, Washington.*

A.D. It was during this interlude of milder weather that the Viking colonization of Iceland and Greenland occurred.

After about 1200 A.D., climatic conditions began to deteriorate again, as noted by a decline in the culture of the Viking colonies. Apparently, a temporary climatic improvement occurred from about 1400 to 1550 A.D., but then a prolonged period of colder, more severe weather set in over much of Europe and North America. The period from approximately 1550 to 1850 A.D. is often referred to as the "Little Ice Age." Glaciers advanced over much of the Northern Hemisphere, and stormy weather conditions were prevalent.

Following this period of severe weather, temperatures began to rise slowly, increasing perhaps one or two degrees Fahrenheit on average around the Northern Hemisphere. This slow warming appeared to end in the 1940s in Europe and in the 1950s in North America. Since then, temperatures have fallen, on average, somewhat less than one degree. These temperature changes seem small, but they are overall averages. Some regions, especially in northern latitudes, have reported larger variations. Also, it is believed that a long-term change of only a few degrees in the average global temperature could mean the difference between the relatively mild climate that we presently enjoy, and another ice age.

Keep in mind that this reconstruction of climatic history is approximate, based on available, but not always consistent, data and theory. Also, much of the data can be interpreted in different ways. The farther back we go in time, the less certain we can be of either climatic conditions or the length of a climatic period. Only during the past one or two centuries have systematic weather records been kept. But, thanks to other historic records and books, we have a good general knowledge of weather that's at least several hundred years old.

## The Year Without a Summer

During the Little Ice Age weather conditions were severe in both North America and Europe. A particularly cold period occurred from 1812 through 1817, producing some of the coldest summer temperatures ever noted in the Mid-Atlantic states and New England.

The summer of 1816 was exceptionally cold in the northeastern United States, and has been referred to as the "year without a summer." Temperatures in New England averaged one to two degrees below normal in May, three to six degrees below normal in June and July, and two to three degrees below normal in August. The spring had been late and vegetation was severely retarded. Then, during the 6th to the 9th of June, a cold outbreak produced frost as far south as Virginia, with ice reported near Philadelphia. In parts of Vermont, ice formed as much as an inch thick. Crops were wiped out in many areas and thousands of birds froze to death. Virtually all of New England's corn crop was destroyed. Light snow fell in northern New York State and northern New England on June 6th, with heavy snow reported in some sections on the 8th.

More favorable weather the remainder of June encouraged farmers to replant some crops, but another outbreak of cold Canadian air swept across New England near the end of the first week in July killing many of the more tender plants, including corn and squash. Toward the end of the summer, frost occurred over much of New England twice during the last ten days of August and, finally, a severe killing frost ended the growing season September 27. The corn harvest in New England produced only about 10 per cent of the expected crop and most other crop yields were far below normal. An interesting account of this event can be found in *American Weather Stories*.[2]

# CAUSES OF CLIMATIC CHANGE

Many different theories have been advanced to explain changes in the earth's climate. Some are plausible, others seem far-fetched. However, until we can better understand the current workings of our atmosphere and oceans, we will not be able to explain accurately how they have changed and interacted in the past. It seems likely that there's not one single cause of climatic change, but rather a combination of several factors, all of which have some effect on our weather. Major factors which may produce changes in our climate are summarized briefly in the following paragraphs. No attempt is made to discuss these in detail—whole books have been devoted to that subject. A selection of references on the subject may be found in Appendix D.

## Astronomical Factors

Such events as changes in the earth's orbit around the sun and a change in the tilt of the earth's axis may have contributed to long-term climatic changes in the remote past, and may influence our climate in the distant future. Such factors cannot, however, readily explain shorter-period fluctuations, such as have occurred during the past ten thousand years.

## Solar Changes

The total amount of radiation emitted from the sun is generally believed to be constant at present, and, as far as can be determined, has been virtually unchanged for a long period of time. There has been some debate on this point, however, and current measurements of the sun's output are not sufficiently precise to give an exact answer to the question of possible small variations in solar radiation. Shorter-term changes in solar activity, such as the sunspot cycle, have been suggested as possible causes of changes in our weather. As yet, however, there is no completely satisfactory explanation of exactly how these solar cycles actually affect atmospheric conditions on earth.

## Volcanoes

Large volcanic eruptions throw many tons of fine particles into the upper atmosphere and therefore affect the amount of solar energy reaching the earth's surface. There is evidence that violent eruptions can lead to a slight decrease in average surface temperatures over an entire hemisphere for at least several months following the explosion. Several major volcanic eruptions have been associated with the 1812 to 1817 cold period described earlier, as well as with the recent cooling trend during the past two decades. There is, however, considerable scientific debate over the real importance of volcanic eruptions to general climatic changes.

*Volcanic eruptions throw tons of fine particles into the upper atmosphere, and these affect the amount of solar energy reaching the earth's surface. Evidence suggests that violent eruptions can lead to a decrease in average surface temperatures. This is the Volcano Cerro Negro in Nicaragua.*

## Continental Drift

The occurrence of a very slow but steady movement of the earth's land masses is now accepted scientific fact. (For years speculation on this movement was considered almost heresy by the scientific community.) Such movements, over periods of millions of years, obviously would affect climatic conditions, as particular regions moved closer to or farther away from the equator. This factor cannot, however, explain shorter-term periodic changes in our climate.

## Carbon Dioxide

Carbon dioxide ($CO_2$) has a significant effect on the earth's surface temperature. It permits the short-wavelength solar radiation to pass through the atmosphere to the surface, but retards the outgoing longwave terrestrial radiation. This combination of effects produces the "greenhouse effect," which tends to trap heat in the lower part of the atmosphere. Generally, an increase in $CO_2$ is associated with warming of the air near the surface; and recent studies suggest that a doubling of the $CO_2$ content could raise average worldwide temperatures as much as five degrees Fahrenheit. Obviously, some other factors are also at work, since temperatures in recent decades have been falling slowly while the $CO_2$ content of the atmosphere has been slowly increasing. (See the discussion on carbon dioxide in Chapter 3.)

*Industrial emissions are a common source of carbon dioxide in the United States. Atmospheric $CO_2$ is increasing slowly, a trend that could have an impact on the earth's temperatures.*

## MAN-MADE CLIMATIC CHANGES

With man's intensified industrial activity and increasing generation of artificial heat, there is an accompanying increase in concern about possible inadvertent man-made changes in our climate. Carbon dioxide content of the atmosphere has been increasing slowly for over a century, primarily due to the burning of fossil fuels. Such an increase, at least in theory, should produce a general warming of the lower atmosphere. A warming trend from about the mid-1800s until the 1940s fits this theory nicely. However, the subsequent cooling noted over much of the northern hemisphere has sent scientists scurrying for new or revised theories.

Some suggest that another of man's additions to the atmosphere—pollution particles—may counteract increased $CO_2$ by reducing the amount of sunlight reaching the surface—an effect similar to that of volcanic dust. To further complicate the picture, however, under certain conditions, some particulate matter can cause heating of the atmosphere by absorption of solar radiation. Thus, the layer of air in which the particles concentrate may be warmed, while lower layers are cooled.

Another factor is power-generating stations which release heat to the atmosphere and surrounding bodies of water. Although the amount of heat released is small compared to the total solar energy received by the earth, this source of heat energy is increasing rapidly.

It's clear that man's activities affect local weather conditions in some areas. Large metropolitan regions produce a "heat island" effect, where nighttime temperatures in and around the center city are as much as five to ten degrees higher than those in surrounding rural areas. Usually, daytime effects are less obvious because of better mixing of the air. Even large shopping center complexes can affect local temperatures significantly, especially at night.[3] Large industrial complexes may modify rainfall patterns for several miles downwind.[4]

Up to now, these man-made weather effects appear to be restricted to the immediate area of the source of the effect and are not believed to influence larger world-wide weather patterns. Such large-scale effects could, however, possibly occur in the future. Once again, until we better understand the natural factors that cause climatic changes, we cannot predict accurately what man-made factors will do.

## THE NEAR FUTURE

What then is the outlook for the near future—say the next several decades? Any significant changes in climate could have disastrous effects on the world's food supply. In many of the major food producing regions of the world, including the United States, the weather during the middle part of the 20th century has been extremely favorable. Any major change in the climate, therefore, is likely to be a change for the worse. Thus, there is a great interest in attempting to estimate future climate trends.

There is a tendency for those who predict grave changes in our climate in the immediate future to receive the most publicity. Hence, it's not unusual to read of the "rapid deterioration" of our climate and the resulting disastrous effects on food production. We need to keep in mind, however, that short-term changes in our climate are normal, rather than abnormal, and that a succession of years with severe drought or unusual coolness do not necessarily foretell permanent drought or persistant cold. For example, severe drought in the northeastern United States during the 1960s has given way to above normal rainfall in the 1970s, along with two major floods and numerous local flash flooding problems (some of which are due, in part, to man's intrusion on the natural floodplain). In at least some parts of drought-stricken Africa, rains have returned. Likewise, in parts of the Mid-Atlantic States, a succession of three or four mild and relatively snow-free winters in the early 1970s replaced the record cold and snow which plagued the region a decade earlier.

In fact, during the 1969-75 period, the cooling trend that had begun some two decades earlier appeared to have eased, and possibly even reversed itself, in many parts of the United States. Just when some scientists began to suggest that the 20-year cooling cycle might have ended, however, record cold blasted much of the country during the latter part of 1976 and early 1977.[5] Fuel shortages were widespread, causing schools and factories to close. In some areas, rivers froze solid for the first time this century and snow fell farther south in Florida than at any time in recorded history.[6] Vegetable and citrus crops in the Sunshine State were virtually wiped out in some areas. This series of events was followed by a second extremely severe winter in 1977-78 over much of the eastern half of the United States, producing one of the worst occurrences of severe back-to-back winter seasons ever recorded in eastern North America.[7] Thus, for the time being at least, the cooling trend appears to be continuing over much of the Northern Hemisphere.

## Weather's Seldom "Normal"

So climatic trends develop, weaken, reverse direction and reappear. For example, take a look at Allentown, Pennsylvania's long-term trends in annual precipitation, seasonal snowfall, and winter temperatures, graphed on pages 156–157. Also plotted as straight, dashed lines are the normal (1941-70) values for these parameters. It is strikingly clear from these graphs that weather is seldom "normal." Not only do rain, snow and winter temperature vary from year-to-year, there are also short-term trends which persist for up to about 10 years. Winter temperatures, for example, remained well above normal from the late 1940s to the late 1950s, turned sharply colder in the early 1960s, and then were persistently above normal in the early 1970s—until the winter of 1976-77. Similar variations occur at most weather stations.

The truth of the matter is: We simply just do not know what physical factors have caused weather changes in the past and therefore we cannot predict what changes will occur in the future. Unless, and until, we have convincing proof that a permanent long-term change in the climate is underway, the most reasonable estimate of climatic conditions in the near future must be based on what has happened in the recent past.

This, unfortunately, is where the major problem comes in—how far back should we go to determine our true "normal" climate? Officially, normal conditions have been determined by taking averages over a 30-year period (1921-50, 1931-60, etc.). Our current normals are based on the 1941-70 period. It had been assumed that such averages were not only representative of current climatic conditions but could be used to make reasonable estimates of future conditions as well.

## Climatic Change?

However, there is increasing evidence that some type of change may, in fact, be occurring in the earth's climate and that the relatively benign conditions most of us have come to consider as normal have actually been ***abnormal***, when compared to longer-term averages.

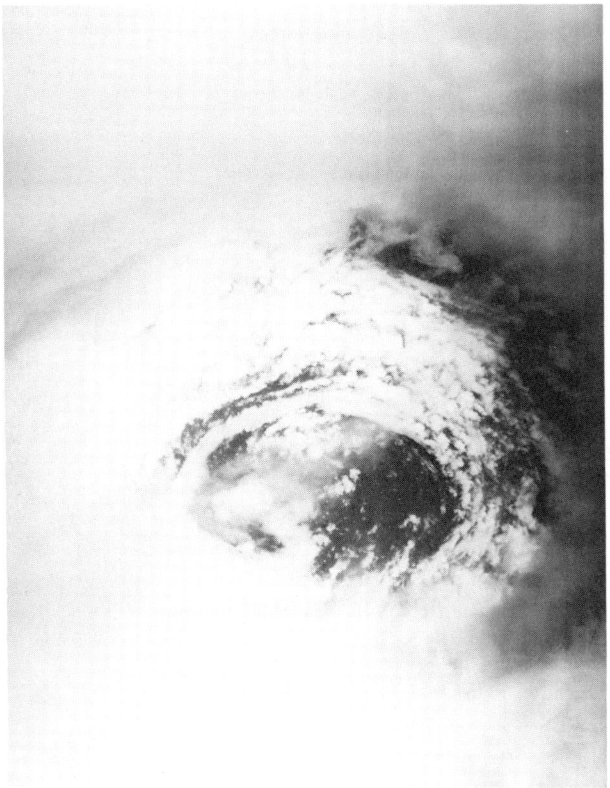

*Aerial photograph of Hurricane Betsy, September 1965.*

Consider, for example, several facts:

1) For only about five to ten per cent of the time during the past million or so years have average temperatures around the globe been as mild as they are now.
2) During the past ten thousand years there have been very few periods as warm as the early and mid-1900s.
3) Evidence suggests that the average interglacial period in the past has lasted about ten thousand years. Our current interglacial period began approximately ten thousand years ago.

Now these facts may sound ominous, and they have been used as the basis for some dire predictions of an upcoming ice age. However, since we don't understand the physical reasons for these past conditions we cannot assume automatically that future changes will exactly repeat those of the past. Nevertheless, until we can provide rational theories to predict future climatic trends, our only realistic guide to upcoming weather conditions remains what has happened in the past.

Also, keep in mind that significant changes in climatic conditions have occurred with only relatively minor changes in average hemispheric temperatures. Average temperatures during the Little Ice Age, for example, were only about three degrees Fahrenheit cooler than those during the peak of the warming trend in the 1940s. Temperature changes were not uniform around the Northern Hemisphere, however, with some areas experiencing changes of perhaps six to eight degrees or more, while other regions showed very slight changes. Nevertheless, these figures are presented to indicate how precariously "balanced" our present climate really is.

## More Weather Instability

Now what does all this mean? I suggest, in view of the data presented above and given the wide variety of weather extremes noted throughout the world in recent years, that our climate is beginning to revert back to a more normal and more unstable state, typical of the past hundreds or even thousands of years. Our so-called "normal" weather, based generally on the decades of the 1920s through the 1960s, was actually not normal at all and gave a false impression of a relatively stable and gentle climate. New evidence of past climatic conditions and current weather extremes suggest that our climate may be entering a period of greater instability and less predictability.

This does not necessarily mean a prolonged period of cooling, although the current trend appears to be in that direction. Past conditions suggest that some further cooling would not be surprising. What the evidence does suggest is a climate where more unusual and extreme weather conditions may become the rule, rather than the exception. These extremes could include heat and drought (such as ravaged parts of Britain and Western Europe in 1976) as well as cold and snow. A more unstable climate, and particularly a cooler one, will likely mean greater storminess. Both floods and droughts may become more numerous.

In general, a picture begins to take shape of a possible change in our climate which, while not necessarily leading to any sudden or dramatic plunge into a new ice age, does point to a climate more hostile and unpredictable (and quite possibly a bit cooler) than what we've become used to in recent decades.

Future years will likely bring detrimental weather to many areas of the globe, including the United States—drought, heat, floods, and freezes. The most serious problems for agriculture are likely to occur where farming is practiced in "fringe" areas—such as on the edge of deserts or in northern latitudes where the growing-season length is critical. There, even slight changes in weather from one year to the next can be disastrous to crop production. However, even gardeners and homesteaders in middle and southern latitudes need to be alert for potential hazards, such as unusually late spring or early fall freezes or prolonged drought conditions. Where and when such events may occur is impossible to forecast, but they seem likely to increase in future years.

In summary, then, there is evidence that the earth's climate is undergoing a subtle but perhaps significant change. A gradual deterioration in climatic conditions, especially as they affect agricultural production, seems possible, since weather conditions during much of the past few decades have been exceptionally good. We could even have another "year without a summer." However, fleeing of northern latitudes to escape the coming ice sheet, or from coastal regions to avoid rising ocean levels, seems premature.

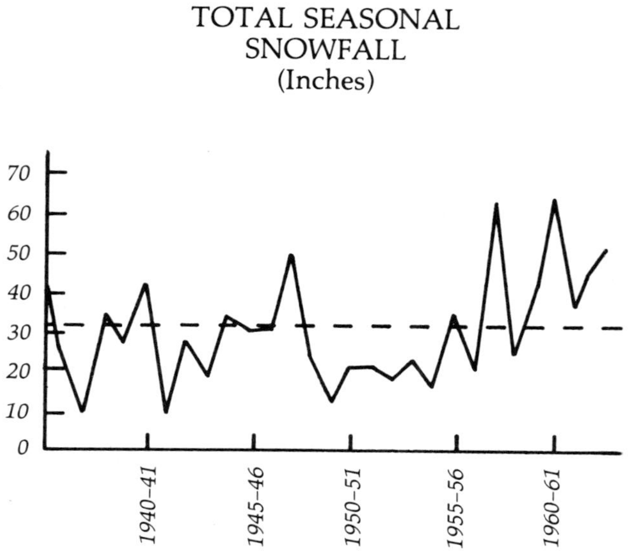


*Long term weather variables at Allentown, Pa.*

# TRACKING YOUR OWN CHANGING CLIMATE

Can you take part in keeping track of our changing climate? You can if you have one or two simple weather instruments described in the first chapter. A max/min thermometer will allow you to follow the ups and downs of temperature trends, while a basic raingage will provide the data to plot monthly and yearly changes in precipitation. Of course you will need a long and continuous record of rainfall and/or temperature to establish any real trends, but even after several years of records you will be able to notice how variable conditions are in your area. You can plot your data on graphs, similar to those shown in the figure *(Long-Term Weather Variables at Allentown, Pa.)* to get a visual picture of how local weather is behaving. Items of particular interest to gardeners and farmers would be the dates of the last spring and first fall freeze and the growing-season rainfall. I've prepared similar graphs,

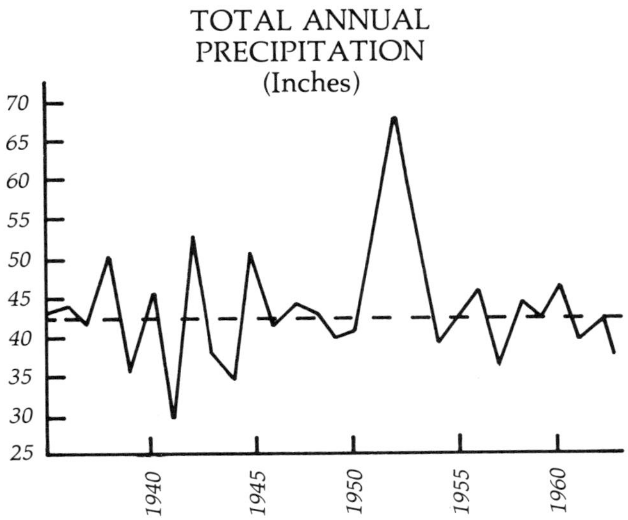


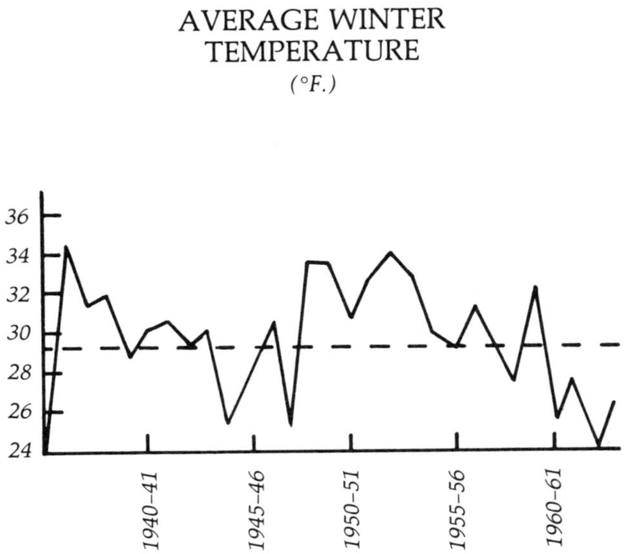


using data recorded at our garden in Camp Hill, Pennsylvania. Is the length of your growing season increasing or decreasing? Does your growing season rainfall seem to be gradually decreasing? You can determine the answers to these questions with your own weather records. A discussion on the analysis of home weather station data is presented in Appendix H.

You can also notice clues to potential changes in climate by closely following whatever phenological data you record (see Chapter 2). The dates on which certain plants leaf, blossom, or set fruit will vary according to weather conditions (mainly temperature) and can be used along with, or in place of, your weather records to determine the variability of the climate in your area. Again, you will need long records to document any permanent change in climate, but as few as a half dozen or so years of data will give you some idea of average conditions for your area. You can also plot these data on graphs to get a better picture of how they vary from year-to-year and to pick out any persistent trends, should they occur.

So pay attention to changes in your local weather and phenological events. They may signal a long-term trend in the overall climate.

# References

## Chapter 2

1. Hacia, H., "Phenology," *Weekly Weather and Crop Bulletin,* Vol. 56, No. 31, 1969, pp. 12-13.

Newman, James E. and Beard, James B., "Phenological Observations: The Dependent Variable in Bioclimatic and Agrometeorological Studies," *Agronomy Journal,* Vol. 54, 1962, pp. 399-403.

Hopp, Richard J., and Blair, B.O., "Plant Phenology in Eastern and Central North America, Part I: Development of Networks and Preliminary Results," University of Vermont Agricultural Experiment Station Bulletin 677, Sept. 1973.

2. Merriam, C. Hart, *"Life Zones and Crop Zones of the United States,"* Bulletin #10, United States Department of Agriculture Division of Biological Surveys, 1898.
3. Hopkins, A. D., "Periodical Events and Natural Law as Guides to Agricultural Research and Practice," *Monthly Weather Review,* Supplement #9, 1918.
4. Hopp, "Charting the Progress of Spring," *Weekly Weather and Crop Bulletin,* Vol. 58, No. 16, 1971, pp. 12-13.
5. Caprio, Joseph M., "Pretty Shrubs and Vines Predict Crop Progress," *Weekly Weather and Crop Bulletin,* Vol. 58, No. 8, 1971, p. 11.
6. Ibid.
7. Newman and Beard, "Phenological Observations," pp. 399-403.
8. Hopp and Blair, "Plant Phenology."
9. Ibid.
10. Fulton, B. B., "The Cabbage Maggot in North Carolina," *North Carolina Agricultural Experiment Station Bulletin 335,* 1942.
11. Whitcomb, W. D. and Garland, W. J., "The Cabbage Maggot and Its Control with Chlorinated Hydrocarbon Insecticides in Massachusetts," *Massachusetts Agricultural Experiment Station Bulletin 540,* 1963.

## Chapter 3

1. Thompson, Louis M., *Weather Variability and the Need for a Food Reserve,* CAED Report 26, Center for Agriculture and Economic Development, Iowa State University, Ames, Iowa, 1966.
2. MacDougal, D.T., "Soil Temperatures and Vegetation," *Monthly Weather Review,* Vol. 31, No. 8, Aug. 1903, pp. 375-379.

3. *How Crops Grow—A Century Later,* ed. by Peter R. Day, Bulletin 708, Connecticut Agricultural Experiment Station, New Haven, Conn. 1969.
4. *Insect and Disease Newsletter No. 296,* College of Agriculture, Rutgers University, New Brunswick, N.J., July 1962.
5. Peek, J.W. and Pendleton, "Light in the Cornfield," *Illinois Research*, Vol. 3, 1965.
6. Palmer, Wayne C., "Keeping Track of Crop Moisture Conditions, Nationwide: The New Crop Moisture Index," *Weatherwise*, Vol. 21, No. 4, August 1968, pp. 156-161.

## Chapter 4

1. Valli, V. J. *Biometeorological Factors Affecting the Development and Spread of Plant Diseases,* NOAA Technical Memo, NWS ER-56, National Weather Service, 1974.
2. Ibid.
3. Scarpa, M. J. and Raniere, Larry C., "The Use of Consecutive Hourly Dewpoints in Forecasting Downy Mildew of Lima Beans," *Plant Disease Reporter,* Vol. 48, number 2, 1964., p. 77-81.
4. Miller, Paul R., "Effect of Environment on Plant Diseases," *Phytoprotection*, Vol. 50, no. 2, 3, 1969, p. 81-94.
5. Walker, John C., *Plant Pathology*, 2nd Ed. McGraw-Hill, New York, 1957.
6. Valli, *Biometeorological Factors.*
7. Rahn, James J. and Barger, Gerald L., "A Summary of Weather Conditions and the Southern Corn Leaf Blight," *Weekly Weather and Crop Bulletin,* Vol. 57, no. 52, Dec. 28, 1970, pp. 11-14.
8. Miller, "Effect of Environment on Plant Diseases."
9. Valli, *Biometeorological Factors.*

## Chapter 5

1. Rodale, Robert, *The Basic Book of Organic Gardening,* Ballantine Books, Inc., New York, N.Y., 1971.
2. Peavy, William S., "Drip Your Way to a Beautiful Garden," *Flower and Garden,* Vol. 21, No. 2, Feb. 1977, p. 35-39.
3. Dickerson, W. H. and Dethier, B. E., *Drought Frequency in the Northeastern United States,* Bulletin 595, West Virginia University Agriculture Experiment Station, 1970.

## Chapter 6

1. Newman, James E. and Dale, Robert F., "Growing Degree Days—Their Use in Forecasting Crop Growth and Maturity," *Weekly Weather and Crop Bulletin,* Vol. 56, no. 32, August 11, 1969, p. 12-13.
2. Ibid.

Gilmore, E. C., Jr., and Rogers, J. S., "Heat Units as a Method of Measuring Maturity in Corn," *Agronomy Journal,* Vol. 50, 1958, pp. 611-615.

Rahn, James J., "Seed Corn Representatives Accept GDD System," *Weekly Weather and Crop Bulletin,* Vol. 57, no. 14, April 6, 1970, p. 15.

3. Caprio, Joseph M., "The Solar Thermal Unit Theory in Relation to Plant Development and Potential Evaporation," Montana Agricultural Experiment Station Circular #251, Montana State University, 1971.
4. Anstey, T. H., "Prediction of Full Bloom Date for Apple, Pear, Cherry, Peach, and Apricot from Air Temperature Data," *Proceedings American Society of Horticultural Science,* Vol. 88, 1966, p. 57-66.

## Chapter 9

1. Leffler, Robert J. and Foster, James L. "A Distinct Precipitation Shadow in the Valley of the South Branch of the Potomac River, West Va.," *National Weather Digest,* Vol. 2, No. 4, November 1977, pp. 21-24.

## Chapter 11

1. Shelton, Marlyn L., "Drought to Flood—A Sudden Moisture Reversal in California," *Weatherwise,* Vol. 31, No. 3, June 1978, pp. 92-100.
2. Hughes, Patrick, *American Weather Stories,* Environmental Data Service, National Oceanic and Atmospheric Administration, U.S. Department of Commerce, 1976 (Available through Superintendent of Documents, Government Printing Office, Washington, D.C. 20402).
3. Martin, Frank P., and Evans, Paul M., "The Heat Island Effect of a Large Shopping Mall in Akron, Ohio," *Weatherwise,* Vol. 28, No. 6, Dec. 1975, p. 254.
4. Huff, F.A., and Chagnon, Stanley A., Jr., "Precipitation Modification by Major Urban Areas," *Bulletin, American Meteorological Society,* Vol. 54, Dec. 1973, pp. 1220-1232.

5. Wagner, A. James, "The Record-Breaking Winter of 1976-77," *Weatherwise*, Vol. 30, No. 2, April 1977, pp. 65-69.
6. Schwartz, Glenn E., "The Day It Snowed in Miami," *Weatherwise*, Vol. 30, No. 2, April 1977, p. 50.
7. Wagner, A. James, "The Cold and Snowy Winter of 1977-78," *Weatherwise*, Vol. 31, No. 2, April 1978, pp. 53-59.

# PHOTO AND ILLUSTRATION CREDITS

Part I, Frederick W. Stetson, Garden Way Publishing Co., Charlotte, Vermont.

CHAPTER 1

p. 1, William Hall, Chittenden County Forester, Chittenden County, Vermont.
p. 2, United States Department of Agriculture.
p. 4, (top), Science Associates, Inc., Princeton, New Jersey
p. 4, (bottom), Taylor Instruments, Arden, North Carolina.
p. 5, Taylor Instruments.
p. 6, Taylor Instruments.
p. 7, (left and right), Science Associates, Inc.
p. 8, (left), Taylor Instruments.
p. 8, (right), Science Associates, Inc.
p. 9, Science Associates, Inc.
p. 10, Taylor Instruments.
p. 11, (left, right and top), Taylor Instruments
p. 12, Science Associates, Inc.
p. 13, Aldren A. Watson, Brattleboro, Vermont.
p. 15, Watson
p. 19, Library of Congress, Washington, D. C.

CHAPTER 2

p. 21, Gardens For All, Inc., Burlington, Vermont.
p. 22, Joseph M. Caprio, Montana Agricultural Experiment Station, Montana State University, Bozeman, Montana.
p. 24, M. T. Vittum, New York State Agricultural Experiment Station, Geneva, N. Y., and Richard J. Hopp, Vermont State Agricultural Experiment Station, Burlington, Vermont.
p. 25, Gardens For All, Inc.
p. 26, Caprio
p. 28, Watson
p. 29, Hopp

Part II, Grant Heipman, Lititz, Pennsylvania.

CHAPTER 3

p. 33, Gardens For All, Inc.
p. 34, Watson

p. 35, Watson
p. 36, Watson
p. 39, Gordon Baer, USDA.
p. 41, Watson
p. 47, Stetson
p. 48, Doug Wilson, USDA.
p. 50, Cathy Baker, Starksboro, Vermont.
p. 54, Geological Survey, U.S. Department of the Interior.

CHAPTER 4

p. 57, USDA
p. 60, Gardens For All, Inc.
p. 63, Environmental Data Service, Environmental Science Services Administration, U.S. Department of Commerce.

CHAPTER 5

p. 67, H.S. Holmes, U.S. Bureau of Reclamation.
p. 69, The National Center For Atmospheric Research, Boulder, Colorado.
p. 70, Peter Clegg, Upper Wraxall, England.
p. 73, Win Way, University of Vermont Extension Agronomist, Burlington, Vermont.
p. 74, Watson
p. 76, USDA
p. 77, B.C. Beville, USDA
p. 81, USDA
p. 82, 83, Watson
p. 84, Gardens For All, Inc.
p. 85, U.S. Bureau of Reclamation.
p. 86, USDA
p. 87, John White, USDA.
p. 88, *The Mother Earth News,* Hendersonville, North Carolina.
p. 89, Erwin W. Cole, Soil Conservation Service, USDA.
p. 90, Joseph Larson, USDA.
p. 91, The National Center for Atmospheric Research.

Part III, Stetson

CHAPTER 6

p. 95, Donald Hanson, Burlington, Vermont.
p. 97, Gardens For All, Inc.
p. 101, Horticultural Research Center, Agricultural Experiment Station, University of Vermont, Burlington, Vermont.
p. 102, Watson
p. 107, U.S. Department of Commerce.

CHAPTER 7

p. 109, Stetson
p. 110, Randy Emmons, Naval Photographic Center, Naval Station, Washington, D.C.
p. p. 11, A.G. Griffith, Naval Photographic Center.
p. 113, George Roska, Naval Photographic Center.
p. 114, Jim Falk, Naval Photographic Center.
p. 115, Stetson
p. 116, National Oceanic and Atmospheric Administration, Washington, D.C.
p. 117, National Oceanic and Atmospheric Administration.

CHAPTER 8

p. 119, National Oceanic and Atmospheric Administration.
p. 121, Straight Arrow Books, San Francisco, California.
p. 122, U.S. Department of Commerce.
p. 123, Pennsylvania Department of Commerce (adapted).
p. 124, National Center For Atmospheric Research.
p. 125, U.S. Department of Commerce (adapted).

CHAPTER 9

p. 129, Watson
p. 130, U.S. Air Force.
p. 134, Watson

CHAPTER 10

p. 137, Vermont Development Agency, Montpelier, Vermont.
p. 138, Gardens For All, Inc.
p. 139, Vermont Development Agency.
p. 140, Watson
p. 142, Baker

CHAPTER 11

p. 143, R.W. Hufnagle, USDA.
p. 145, Naval Photographic Center
p. 147, U.S. Geological Survey, Department of The Interior.
p. 148, U.S. Geological Survey.
p. 150, Naval Photographic Center.
p. 151, U.S. Geological Survey.
p. 152, Stu Perry, The Burlington Free Press, Burlington, Vermont.
p. 155, U.S. Air Force
Appendix, Cal Olson, USDA

# Appendix

# A
# GLOSSARY

ADVECTION FREEZE - a freeze caused by the movement of a mass of cold air, with temperatures 32° F. or lower, over an area.

AGRICULTURAL CLIMATOLOGY - the study of the effects of climate on agricultural activities; the application of climatological principles to agriculture.

AGRICULTURAL DROUGHT - the lack of sufficient moisture in the soil to maintain adequate growth and development of crops.

ANEMOMETER - an instrument to measure the speed of the wind.

AVAILABLE WATER - the amount of moisture in the soil that is actually available for use by plants; it is the amount of water held in the soil between the field capacity and the permanent wilting point.

BAROGRAPH - a recording barometer.

BAROMETER - an instrument used to measure air pressure; there are two basic types, aneroid and mercury.

BASE TEMPERATURE - in calculating growing degree days (heat units) it is the lowest temperature at which growth and development of the plant occurs; the accumulating of heat units begins at this base temperature.

BLACK FROST - the occurrence of a killing frost without the appearance of frost crystals (hoarfrost); sometimes referred to as a dry freeze.

CALM - lack of any apparent movement of the air.

CARBON DIOXIDE - a colorless gas, comprising about 0.033 per cent of the total atmosphere; carbon dioxide ($CO_2$) combines with water in the photosynthetic process to form new plant material; in the atmosphere it retards the loss of heat by longwave radiation from the earth's surface, producing the so-called "greenhouse effect."

CARDINAL TEMPERATURES - critical temperatures in the growth and development of plants; these include the minimum, optimum, maximum, and lethal temperatures.

CHILL INJURY - injury caused to plants by low, but not freezing, temperatures.

CHLOROPLASTS - tiny microscopic bodies within plant cells in which the process of photosynthesis is centered.

CHLOROPHYL - substance found within chloroplasts which is capable of absorbing certain wavelengths of solar radiation and using this energy to promote the photosynthetic process.

CLIMATE - the long-term atmospheric conditions of an area; the "average" weather conditions, along with normal variations and extremes.

CLIMATIC OPTIMUM - an extremely favorable period of mild climatic conditions during interglacial periods; especially applied to period around 5000 to 2500 BC.

CLIMATOLOGY - the study of climate, including analyses of climatic data.

CLOUD COVER - that portion of the sky covered by clouds, usually measured in tenths.

COLD AIR DRAINAGE - the downslope movement of cold, dense air.

COLD FRONT - the leading edge of a mass of cool or cold air that is replacing a mass of milder air.

CROP MOISTURE INDEX (CMI) - an index designed to measure large-scale soil moisture conditions as they affect agricultural production.

DEW - droplets of water which have condensed on objects, such as plants, which have cooled below the dewpoint temperature of the air.

DEWPOINT TEMPERATURE - the temperature to which a given parcel of air must be cooled before it becomes saturated; the temperature of an object when dew first forms on it.

DIURNAL - daily; refers to events which recur every 24 hours, such as a daily temperature cycle.

DROUGHT - a long-term water shortage as applied to agricultural needs, water-supply reserves, streamflows, etc.; two general types are recognized—agricultural drought and meteorological drought.

DROUGHT INDEX - also known as the Palmer Index; a measure of the overall water deficit as reflected in stream flow, reservoir levels, groundwater tables, etc.

EVAPORATION - the loss of water into the atmosphere by its conversion from a liquid to a gaseous state.

EVAPORIMETER - an instrument to measure the loss of water by evaporation.

EVAPOTRANSPIRATION - total loss of water from a crop canopy due to evaporation plus transpiration.

FIELD CAPACITY - maximum amount of water held by the soil after surface water has drained off and water lost by gravity is removed.

FREEZE - the occurrence of air temperatures of 32° F. (0°C.) or lower.

FREEZE-FREE SEASON - period between last 32° F. in spring and first 32° F. in fall.

FREEZE PROBABILITIES - the probability (risk) of a freeze occurring after a given date in the spring, or before a given date in the fall.

FROST - also called white frost, or hoarfrost; the whitish covering of ice crystals found on objects when their surface temperature has fallen to 32° F. or lower.

FROST POCKET - a pool of cold air, usually at least several degrees colder than surrounding areas; found mainly in low-lying areas of uneven terrain.

GLACIERS - large masses of land ice; may be slowly advancing or receding.

GREENHOUSE EFFECT - the tendency for certain gases in the atmosphere, especially water vapor and carbon dioxide, to absorb some of the earth's outgoing longwave radiation and reflect it back to earth; this prevents surface temperatures from cooling as much as they otherwise would and is analogous to the effect of glass used in a greenhouse, hence the name.

GROWING DEGREE DAYS - an accumulation of daily mean temperatures above a certain base value; used as a measure of crop growth and development.

GROWING SEASON - period between the last killing frost in spring and first killing frost in fall; may be shorter than freeze-free season (see earlier definition).

HARDENING-OFF - the process whereby a plant gradually builds up a resistance to adverse environmental conditions, especially cold.

HARDINESS - a measure of a plant's ability to withstand cold or other weather stresses.

HEATING DEGREE DAYS - an accumulation of daily mean temperatures below a fixed base temperature (usually 65° F.) which is used as a measure of fuel consumption required to maintain a given amount of warmth in buildings.

HEAT ISLAND - an area over which air temperatures are higher than those over surrounding regions; usually used in describing the increased warmth generally found in metropolitan areas, especially at night.

HEAT UNIT - a measure of temperature that can be accumulated over a period of time; degree days, both growing and heating, are examples of heat units.

HOARFROST - see Frost.

HOST PLANT - a plant infected by a disease organism.

HUMIDITY - a measure of the amount of water vapor in the air.

HYDROTHERMOGRAPH - a recording instrument which provides continuous graphs of both temperature and relative humidity values.

ICE AGE - a period in the earth's history during which glaciers covered much of the land area of temperate and high-latitude regions.

ICE PELLETS - the official name for the frozen precipitation popularly called sleet.

INCIPIENT WILTING - the occurrence of a moisture deficit within a plant which reduces the water content of the plant's cells but is not yet severe enough to cause visible, permanent wilting; a temporary wilting, usually occurring during warm, dry and sunny days when the rate of evapotranspiration exceeds the amount of water taken up by the plant's root system.

INDICATOR PLANT - a plant used to make phenological observations of its rate of growth and development.

INSTRUMENT SHELTER - a structure designed to hold certain weather instruments, particularly thermometers, to protect them from direct exposure to sunlight, precipitation, etc.

INTERGLACIAL PERIOD - an interval of time between glacial periods when climatic conditions are milder and generally more benign.

INVERSION - a condition in which the temperature of the air near the surface increases with height; inversions may extend upwards several hundred or even several thousands of feet.

IRRIGATION - the artificial application of water to a garden or cropland to prevent loss of crops due to lack of rainfall.

LAKE BREEZE - a pattern of wind flow set in motion by the difference in temperature between a body of water and the surrounding land mass.

LANGLEY - a measure of radiation, including solar radiation; one langley equals one calorie per square centimeter.

LETHAL TEMPERATURES - those temperatures, either hot or cold, which result in the death of a plant.

LIMITING FACTORS - in plant growth, those factors which act to prevent a plant from achieving its maximum possible rate of growth; these may include sunlight, moisture, carbon dioxide, etc.

LITTLE ICE AGE - the period from approximately 1550 to 1850 AD when temperatures in Europe and North America averaged several degrees colder than at present, glaciers advanced in many areas, and weather conditions were more extreme.

MACROCLIMATE - the general climate over large areas, such as Eastern North America, Western Europe, etc.

MATURITY RATING - the rating of a particular plant variety in terms of the length of time required from planting until it is ready for harvest; usually given in numbers of calendar days or growing degree days.

MAXIMUM TEMPERATURE - the highest temperature recorded during a specified period, usually 24 hours.

MAXIMUM/MINIMUM THERMOMETER - an instrument which indicates the highest and lowest temperature during a specified period.

MEAN TEMPERATURE - the average of the maximum and minimum temperatures for a particular period; the mean equals the sum of the max and min divided by two.

MESOCLIMATE - the climate of localized areas, such as in and around a large city, or around a large body of water.

METEOROLOGICAL DROUGHT - an extended period of dryness which affects streamflow, ground-water tables, and reservoir levels.

METEOROLOGY - the study of atmospheric conditions and the resultant weather they produce; the science of weather analysis and forecasting.

MICROCLIMATE - the localized and unique climate found near the surface of the earth and restricted to small areas, such as a backyard garden, or surrounding a small pond.

MINIMUM TEMPERATURE - the lowest temperature observed during a specified period, usually 24 hours.

MOISTURE STRESS - the adverse effect on plant development caused by lack of a sufficient amount of moisture.

NORMAL - in climatology, "normal" values are defined as averages taken over a 30-year period and generally updated every ten years; in popular usage, normal and average are often used interchangeably.

OPTIMUM TEMPERATURE - the temperature, or temperature range, at which plant growth and development proceeds at its greatest rate.

OROGRAPHIC PRECIPITATION- rain or snow caused by the lifting of an air mass over a mountain, or mountain range; the lifting causes cooling of the air and condensation of the water vapor in the air mass.

PATHOGEN - the organism, or agent, which causes plant disease.

PERMANENT WILTING POINT - the moisture content of the soil below which irreversible wilting of a crop occurs.

PHENOLOGICAL EVENT - a particular happening in a plant's normal course of development, such as leafing, blossoming, or fruiting.

PHENOLOGY - the study of natural periodic events in the life-cycle of plants (or animals).

PHOTOPERIODISM - the effect of daylength on the development of plants.

PHOTOSYNTHESIS - the process whereby plants produce new plant material from the basic raw materials carbon dioxide and water in the presence of sunlight; in this process, solar energy is used by the plants to split water molecules and force a combination with $CO_2$ to produce organic molecules.

PHOTOTROPISM - the effect of light on the direction of growth of a plant.

PLANT DEVELOPMENT - the progress from one stage to another in the life-cycle of a plant.

PLANT GROWTH - the increase in size, weight, or volume of a given plant or crop.

PRECIPITATION - any form of water that falls from the atmosphere, either as a liquid or a solid, and reaches the ground; this includes rain, snow, sleet, hail, etc.

PRECIPITATION PROBABILITY - the chance that a measureable amount of precipitation will fall at a given point during a specified time period.

PSYCHROMETER - an instrument used for measuring the water vapor content of the air.

PYRHELIOMETER - an instrument used to measure direct solar radiation.

RADIATION - a form of energy transmission, usually envisioned as occurring in a wave motion.

RADIATION FROST - a frost which occurs due to radiational cooling of plant and other surfaces; occurs with clear skies and very light or calm winds; usually the last frosts in spring and the first frosts in fall are of this type.

RADIATION SPECTRUM - the complete range of radiation wavelengths, from very shortwave cosmic rays, through the visible wavelengths and into the longwave radio waves.

RADIATIONAL COOLING - the loss of heat by the earth's surface and adjacent layers of air by longwave radiation into space; most pronounced on clear, calm nights.

RAINGAGE - an instrument used to measure rainfall (or frozen precipitation).

RAIN SHADOW - an area of reduced precipitation found downwind from a large mountain or mountain range which acts as a barrier to the general windflow; most of the moisture in the air is deposited as precipitation on the upwind side of the mountain and the air drys out as it flows down the other side.

RECORDING RAINGAGE - a raingage in which the amount of precipitation is recorded on a chart, producing a permanent and continuous record.

RELATIVE HUMIDITY - the amount of moisture in the air compared to the maximum amount of moisture the air could hold at a given temperature; measured as a percentage of saturation.

RESPIRATION - the biological process in plants whereby plant material is broken down and energy is released; in a sense it is the reverse of photosynthesis.

RUNOFF - water from precipitation that is unable to penetrate into the soil layer and flows off the land and into streams and drainage ditches; the water may not infiltrate the soil because of the intensity of the rainfall, soil that is already saturated, or a hard surface layer; can cause considerable erosion.

SEA BREEZE - see Lake Breeze.

SHELTERBELT - a planting of trees and/or shrubs designed to reduce the windflow over a particular land area; may be used to reduce wind damage, evapotranspiration, or other undesirable effects of excessive air movement.

SIX'S THERMOMETER - one type of maximum/minimum thermometer in which a U-shaped glass tube is partially filled with mercury, which moves metal markers within the tube to indicate the highest and lowest temperatures since the last setting.

SLING PSYCHROMETER - a type of psychrometer in which two thermometers, one covered by a wet muslin wick, are swung through the air by hand until both instruments reach their lowest reading. This produces a wet-bulb and a dry-bulb temperature reading which can then be converted into values for the dewpoint and the relative humidity.

SOIL MOISTURE - the amount of water held within the soil; in agriculture it usually refers to moisture within the root zone of crop plants.

SOIL THERMOMETER - an instrument used to obtain the temperature of the soil at any given depth.

SOLAR ENERGY - see Solar Radiation.

SOLAR RADIATION - the radiation (energy) emitted by the sun and covering all wavelengths from ultraviolet to infrared, including the visible "sunlight."

SPECTRUM - see Radiation Spectrum.

STATIONARY FRONT - a boundary line separating warm and cold air which shows little or no movement; usually the result of a cold or warm front that has stalled.

STOMATES - tiny pore-like openings on plant leaves through which various gasses, such as oxygen and carbon dioxide, and water vapor move.

TEMPERATURE GRADIENT - the change in temperature, either vertical or horizontal, from one point to another.

TEMPERATURE INVERSION - see Inversion.

THERMOGRAPH - a recording thermometer.

THERMOMETER SHELTER - see Instrument Shelter.

TRANSPIRATION - the loss of water vapor by plants through stomates; moisture within the plant evaporates and moves through the intercellular spaces to the stomates, through which it passes into the atmosphere.

VECTOR - an insect which acts as a carrier for a virus disease of plants.

VERNALIZATION - the exposure of seeds or seedlings to a period of low temperatures to speed-up, or by-pass, certain parts of the normal life cycle of the plant; used to transform winter grains into spring varieties.

VISIBLE SPECTRUM - that part of the radiation spectrum visible ot the human eye.

WARM FRONT - the leading edge of a mass of warm air which is displacing cooler air.

WARNINGS - issued to alert the public that potentially damaging weather conditions (flash floods, heavy snows, etc.) are already occurring or are imminent.

WATCHES - issued when dangerous weather conditions are possible but not yet certain.

WATER BALANCE - the net effect on soil mositure of rainfall, evaporation, water uptake by plants, and other water losses; the balance between water gain and water loss.

WAVELENGTH - the distance between wave crests; used as a description of radiation type (shortwave or longwave).

WEATHER - the state of the atmosphere at any given time or over a short period of time; the day-to-day changes in atmospheric conditions.

WEATHER SHELTER - see Instrument Shelter.

WETTING PERIOD - the time when leaf or other plant surfaces are covered by liquid water; used mainly when describing conditions related to plant diseases.

WINDBREAK - see Shelterbelt.

WIND DIRECTION - the direction *from* which the wind is blowing.

WIND MACHINE - large fan-like device used to help prevent frost by mixing milder air aloft with cold air near the surface.

WINTER KILL - death of plants caused by the inability of the root systems to absorb sufficient moisture from frozen or frost-heaved soils to compensate for water lost by transpiration; most common with evergreens which continue to transpire on sunny, relatively mild and often breezy winter days.

# B
# SUPPLIERS OF WEATHER INSTRUMENTS

While regular thermometers can be purchased at a variety of stores, more specialized weather equipment usually is available only through particular suppliers. Some basic equipment, such as maximum/minimum thermometers, soil thermometers, and small plastic raingages may often be found at garden centers and occasionally at specialty discount stores. A few seed catalogs also advertise some basic weather instruments. As with anything else, you usually get what you pay for and a few extra dollars invested on a good quality instrument will be money well spent.

Buying directly at a garden center or similar specialty store gives you the advantage of seeing the merchandise and perhaps comparing several models before you make your purchase. Also, it is generally easier and quicker to return faulty items to a local store then to repackage them and mail them back to the place of purchase.

However, not all areas may have a nearby store that carries basic weather instruments and once you get past the standard thermometers and raingages your best (and perhaps only) source for good quality weather instruments is one of the manufacturers specializing in such equipment. The following list of suppliers is certainly not all-inclusive, but does include some of the better known companies that handle the basic type of weather instruments as well as more elaborate equipment. All of the companies listed below answered a postcard request for one of their catalogs and generally carry good quality instruments. By looking over catalogs from two or three of these dealers you will get a good comparison of the variety and prices of equipment available.

Science Associates, Inc.
Box 230
230 Nassau St.
Princeton, N.J. 08540

Belfort Instrument Co.
1600 S. Clinton St.
Baltimore, Md. 21224

Don Kent Instruments
283 Washington St.
Weymouth, Mass. 02188

EPIC, Inc.
150 Nassau St.
New York, N.Y. 10038

C. W. Thornthwaite Assoc.
Route 1, Centerton
Elmer, N.J. 08318

Weatherwise Books
& Instruments
Main Street
New London, N.H. 03257

Weksler Instruments Corp.
Freeport, N.Y. 11520

Taylor Instruments
Consumer Products Divisio
Sybron Corp.
Arden, North Carolina 2870

WeatherMeasure Corp.
P.O. Box 41257
Sacramento, Calif. 95841

# C
# SOURCES OF WEATHER AND CLIMATE DATA

A tremendous amount of weather and climate data has been accumulated during recent decades. Much of this data has been processed in one form or another and can be readily obtained by the general public. Climatological data of particular interest to gardeners and farmers include long-term growing season rainfall and temperature averages, mean frost dates in spring and fall, growing season lengths, etc. Such summaries are the best sources of climatological information for those persons without their own weather station. Even for those of you who keep your own records, the "official" data offer an interesting comparison. In addition to summaries of long-term averages, "raw" weather data (actual daily or monthly values of temperature, rainfall, etc., for a particular year) are also available. Some of these data (either "raw" or summarized) may be obtained from the nearest weather station or agricultural extension office, while others must be ordered through government sources.

## National Sources

At one time, the National Weather Service (formerly the U.S. Weather Bureau) operated a nationwide climatology program in which each state or group of states had its own state climatologist. This person was the climate "expert" and principal source of climatological data for that state. Unfortunately, the National Oceanic and Atmospheric Administration (NOAA), parent agency of the National Weather Service, withdrew its funding of this program in 1973 and the service was terminated. (Some states maintained or re-established a climate program on their own when the federal program ended. Passage of the National Climate Program Act by Congress in late 1978 has made matching federal funds available to states which establish a climate program, so additional states may follow suit.) Virtually all climatological data for the United States (as well as much foreign data) are now collected and archived at the National Climatic Center in Asheville, North Carolina.

Individual National Weather Service (NWS) offices maintain records primarily for their station only and not for other locations, even within the same state. Thus, if you desire weather information for a specific town in which a NWS office is located, you can probably obtain the data directly from that office. If you need data for other areas or for several different regions, you will probably need to contact the National Climatic Center (NCC). A call or letter of inquiry should indicate as plainly and specifically as possible the type of data you require, the location you are interested in, and the period of record you desire. For example, you might request the average monthly rainfall for several locations in eastern Kansas for the past ten years. If you aren't exactly sure just what type of data you need, explain your problem as clearly as possible and ask for any information they might have that would be useful for your requirements. Since there is a charge involved it would be wise to first inquire whether the data you want is readily available and what the cost will be. The address of the NCC is:

National Climatic Center
Federal Building
Asheville, N.C. 28801

## Local Sources

You may be able to acquire the data you want closer to home. As mentioned earlier, local NWS offices will have weather information for their own location, but usually not for other areas. Give them a call or drop them a letter to see if the data you need is available through their office. If not, they will probably refer you to the National Climatic Center. The National Weather Service Office can be located in your telephone book in the "U.S. Government" listing under Commerce Department.

In many states, agricultural experiment stations and/or agricultural colleges have worked up climatic data of particular importance to agricultural interests, such as frost dates, growing season lengths, and growing season rainfall. Usually the data will cover either the entire state or selected regions. Check with your county agent, farm advisor, or agricultural experiment station to see what weather data have been published for your area. Also, if you have some specific weather related problem, they may be able to help you on an individual basis. In many areas, there is a definite need for states to develop programs to provide weather services for agricultural interests, including home gardeners and small farmers.

## Other Sources

Some public libraries, especially larger ones, have a file of climatological information and may receive one or more of the publications listed in the following section. Check with your local library to see if they have, or can obtain, the ones you are interested in. Many college and university libraries will also have these publications. Officials of the nearest National Weather Service office should have available climatological publications covering their state and perhaps some for surrounding states.

## Publications

There are several publications which provide reasonably current weather information. Generally, these are issued monthly, with an annual summary. The following list includes the primary publications likely to be of interest to the general public. The cost quoted is current as of early 1979, but of course is subject to change. In some cases, individual copies are available and you may wish to obtain one to check its usefulness before you order a full subscription.

| *Publication* | *Cost* | *Information* |
|---|---|---|
| **Climatological Data** | $5.10 per year, including annual summary. Single copy 40¢. | Published monthly, by state. Includes daily temperature and precipitation data for all climatological stations in the state, along with monthly averages. July issue includes snowfall and heating degree day data for previous winter. Order by individual state. |
| **Hourly Precipitation Data** | $5.00 per year, including annual summary. Single copy 40¢. | Monthly, by state. Contains hourly and daily rainfall records for each station equipped with a recording raingage. Order by state. |
| **Local Climatological Data** | $2.55 per year, including annual summary. Single copy 20¢. | Monthly. Issued for each National Weather Service office. Contains temperature, precipitation, snowfall, heating degree day and wind data. Annual summary includes monthly values for past years. Order by specific station. |

The latest official climatological normals for temperature, precipitation, and heating and cooling degree days, based on the 1941-70 period are available for individual states. When ordering, ask for the *Climatography of the United States No. 81: Monthly Normals of Temperature, Precipitation, and Heating and Cooling Degree Days 1941-70,* for each state you are interested in. Cost is only 25¢ per state.

All of the above publications should be ordered from:

National Climatic Center
Federal Building
Asheville, N.C. 28801
Attn: Publications

You can also write to the National Climatic Center and request a complete list of all their current climatological publications.

Another publication which might be of interest to gardeners and farmers is the national *Weekly Weather and Crop Bulletin.* Subscription price is $13.00 per year (52 issues). This publication contains weekly temperature and precipitation data for a select list of United States cities as well as monthly summaries. Brief weekly summaries of weather conditions and agricultural activities are included for each state as well as a National Summary. Periodic summaries of world-wide weather conditions as they affect agriculture are presented. Growing degree day and Crop Moisture Index Maps are published during the growing season and heating degree day data are included during the winter. This informative publication should be ordered from:

Agricultural Weather Facility
Room 1248, South Bldg.
U.S. Dept. Agriculture
Washington, D.C. 20250

In addition, many states publish their own weekly crop and weather reports, at least during the growing season. These are often available free of charge or can be obtained for a nominal fee. Check with your county agent or state department of agriculture to inquire about getting on their mailing list.

## Popular Journals

There is really only one popular-style periodical published primarily for amateur weathermen and anyone interested in weather and its effects on human activities. The magazine is called

*Weatherwise* and is published bi-monthly. Each issue contains several feature articles written for the layman, special features on weather items of unusual or historical interest, and a summary of national weather conditions for a recent two-month period. Brief weather summaries are included for the Canadian provinces. I believe that most persons interested in weather on a popular level will find this little magazine appealing. Subscription information can be obtained by writing to:

Weatherwise
Heldref Publications
4000 Albemarle St., N.W.
Washington, D.C. 20016

# NOAA Weather Radio Transmitters In Operation As Of December 1, 1978

The number following each city identifies the radio frequency on which the station transmits. They are:

(1) 162.550 MHZ
(2) 162.400 MHZ
(3) 162.475 MHZ

**Alabama**
Anniston ....................(3)
Birmingham ....................(1)
Dozier ....................(1)
Florence ....................(3)
Huntsville ....................(2)
Louisville ....................(3)
Mobile ....................(1)
Montgomery ....................(2)
Tuscaloosa ....................(2)

**Alaska**
Anchorage ....................(1)
Cordova ....................(1)
Fairbanks ....................(1)
Homer ....................(2)
Juneau ....................(1)
Ketchikan ....................(1)
Kodiak ....................(1)
Nome ....................(1)
Petersburg ....................(1)
Seward ....................(1)
Sitka ....................(1)
Valdez ....................(1)
Wrangell ....................(2)
Yakutat ....................(1)

**Arizona**
Phoenix ....................(1)
Tucson ....................(2)

**Arkansas**
Fayetteville ....................(3)
Fort Smith ....................(2)
Gurdon ....................(3)
Jonesboro ....................(1)
Little Rock ....................(1)
Star City ....................(2)
Texarkana ....................(1)

**California**
Coachella ....................(2)
Crescent City, Calif.
Brookings, Oreg. ....................(1)
Eureka ....................(2)
Fresno ....................(2)
Los Angeles ....................(1)
Monterey ....................(2)
Point Arena ....................(2)
Sacramento ....................(2)
San Diego ....................(2)
San Francisco ....................(1)
San Luis Obispo ....................(1)
Santa Barbara ....................(2)

**Colorado**
Denver ....................(1)

**Connecticut**
Hartford ....................(3)
Meriden ....................(2)
New London ....................(1)

**Delaware**
Lewes ....................(1)

**District of Columbia**
Washington, D.C. ....................(1)

**Florida**
Fort Myers ....................(3)
Daytona Beach ....................(2)
Gainesville ....................(3)
Jacksonville ....................(1)
Key West ....................(2)
Melbourne ....................(1)
Miami ....................(1)
Orlando ....................(3)
Panama City ....................(1)
Pensacola ....................(2)
Tallahassee ....................(2)
Tampa ....................(1)
West Palm Beach ....................(2)

**Georgia**
Atlanta ....................(1)
Athens ....................(2)
Augusta ....................(1)
Chatsworth ....................(2)
Columbus ....................(2)
Pelham ....................(1)
Savannah ....................(2)
Waycross ....................(3)

**Hawaii**
Hilo ....................(1)
Honolulu ....................(1)
Kokee ....................(2)
Mt. Haleakala ....................(2)

**Idaho**
Boise ....................(1)
Lewiston ....................(1)
Pocatello ....................(1)

**Illinois**
Chicago ....................(1)
Moline ....................(1)
Peoria ....................(3)
Rockford ....................(3)
Springfield ....................(2)

**Indiana**
Evansville ....................(1)
Fort Wayne ....................(1)
Indianapolis ....................(1)
Lafayette ....................(3)
South Bend ....................(2)

**Iowa**
Des Moines ....................(1)

**Kansas**
Colby ....................(3)
Concordia ....................(1)
Dodge City ....................(3)
Topeka ....................(3)
Wichita ....................(1)

**Kentucky**
Ashland ....................(1)
Bowling Green ....................(2)
Covington ....................(1)
Hazard ....................(3)
Lexington ....................(2)
Louisville ....................(3)
Mayfield ....................(3)
Somerset ....................(1)

**Louisiana**
Baton Rouge ....................(2)
Lake Charles ....................(2)
Morgan City ....................(3)
New Orleans ....................(1)
Monroe ....................(1)
Shreveport ....................(2)

**Maine**
Ellsworth ....................(2)
Portland ....................(1)

**Maryland**
Baltimore ....................(2)
Salisbury ....................(2)

**Massachusetts**
Boston ....................(2)
Hyannis ....................(1)

**Michigan**
Alpena ........................(1)
Detroit ........................(1)
Flint ........................(2)
Grand Rapids ........................(1)
Marquette ........................(1)
Sauit Ste. Marie ........................(1)
Traverse City ........................(2)

**Minnesota**
Duluth ........................(1)
Int'l. Falls ........................(1)
Mankato ........................(2)
Minneapolis ........................(1)
Rochester ........................(3)
St. Cloud ........................(3)
Thief River Falls ........................(1)
Willmar ........................(2)

**Mississippi**
Ackerman ........................(3)
Booneville ........................(1)
Bude ........................(1)
Gulfport ........................(2)
Inverness ........................(1)
Jackson ........................(2)
McHenry ........................(3)
Meridian ........................(1)
Oxford ........................(2)

**Missouri**
Camdenton ........................(1)
Hannibal ........................(3)
Kansas City ........................(1)
St Joseph ........................(2)
St. Louis ........................(1)
Springfield ........................(2)

**Montana**
Billings ........................(1)
Glasgow ........................(1)
Great Falls ........................(1)
Helena ........................(2)

**Nebraska**
Omaha ........................(2)

**Nevada**
Reno ........................(1)
Winnemucca ........................(2)

**New Hampshire**
Concord ........................(3)

**New Jersey**
Atlantic City ........................(2)

**New Mexico**
Albuquerque ........................(2)
Clovis ........................(3)
Farmington ........................(3)
Hobbs ........................(2)
Ruidoso ........................(1)
Santa Fe ........................(1)

**New York**
Binghamton ........................(3)
Buffalo ........................(1)
New York City ........................(1)
Rochester ........................(2)
Syracuse ........................(1)

**North Carolina**
Cape Hatteras ........................(1)
New Bern ........................(2)
Wilmington ........................(1)

**North Dakota**
Fargo ........................(3)

**Ohio**
Akron ........................(2)
Caldwell ........................(3)
Cleveland ........................(1)
Columbus ........................(1)
Dayton ........................(3)
Lima ........................(2)
Sandusky ........................(2)
Toledo ........................(1)

**Oklahoma**
Lawton ........................(1)
Oklahoma City ........................(2)
Tulsa ........................(1)

**Oregon**
Astoria ........................(2)
Coos Bay ........................(2)
Eugene ........................(2)
Newport ........................(1)
Portland ........................(1)

**Pennsylvania**
Allentown ........................(2)
Erie ........................(2)
Harrisburg ........................(1)
Philadelphia ........................(3)
Pittsburgh ........................(1)
Wilkes-Barre ........................(1)
Willamsport ........................(2)

**Puerto Rico**
San Juan ........................(2)

**Rhode Island**
Providence ........................(2)

**South Carolina**
Beaufort ........................(3)
Charleston ........................(1)
Columbia ........................(2)
Florence ........................(1)
Greenville ........................(1)
Myrtle Beach ........................(2)

**Tennessee**
Bristol ........................(1)
Chattanooga ........................(1)
Knoxville ........................(3)
Memphis ........................(3)
Nashville ........................(1)

**Texas**
Abilene ........................(2)
Amarilla ........................(1)
Austin ........................(2)
Big Spring ........................(3)
Borwnsville ........................(1)
Bryan ........................(1)
Corpus Christi ........................(1)
Dallas ........................(2)
Del Rio ........................(2)
Fort Worth ........................(1)
Galveston ........................(1)
Houston ........................(2)
Laredo ........................(3)
Lubbock ........................(2)
Lufkin ........................(1)
Midland ........................(2)
Paris ........................(1)
Pharr ........................(2)
San Antonio ........................(1)
Victoria ........................(2)
Wichita Falls ........................(3)
San Angelo ........................(1)
Sherman ........................(3)
Tyler ........................(3)
Waco ........................(3)

**Utah**
Salt Lake City ........................(1)

**Vermont**
Burlington ........................(2)

**Virginia**
Norfolk ........................(1)
Richmond ........................(3)

**Washington**
Neah Bay ........................(1)
Seattle ........................(1)
Yakima ........................(1)

**West Virginia**
Charleston ........................(2)
Clarksburg ........................(1)

**Wisconsin**
Green Bay ........................(1)
LaCrosse ........................(1)
Madison ........................(1)
Menomonie ........................(2)
Milwaukee ........................(2)
Wausau ........................(3)

# D
# ADDITIONAL READING MATERIAL

This book has been written especially for backyard gardeners, homesteaders, and small farmers, and for all other laypersons interested in weather and climate and how they affect plant growth. The contents have been kept as non-technical as possible while still providing enough basic background material for the reader to get a general understanding of the subject.

References are listed at the end of some of the chapters for those readers interested in pursuing further a specific item noted in the text. I've tried to keep those references at the popular or semi-popular level, avoiding the more technical articles found in scientific journals.

There are many additional articles and books on the general subject of weather, climate and agriculture, but the vast majority are quite technical, written primarily for other scientists or persons with a strong scientific background. Most of them are not suitable for the general public. There are not many popular articles or books available in the field of agricultural weather, but I've tried to provide at least a few in the following listing. These general references are intended for those readers who may have found one or more topics in the book of particular interest and would like to do some more in-depth reading on those subjects. Sources of more scientific and technical literature are listed in Appendix E.

## Weather Hazards in Agriculture

**Articles:**

"Canadian Cold and Heat Data, "*Weatherwise* Vol. 24, No. 6, Dec. 1971, pp. 288-291.

Lansford, Henry, "The National Hail Research Experiment," *Weatherwise,* Vol. 29, No. 4, Aug. 1976, p.168.

**Reports, booklets, books:**

Boone, L.M., *Estimating Crop Losses Due to Hail,* Agricultural Economic Report 267, U.S. Department Agriculture, Washington, D.C. 1974.

Davis, J. R., "*Frost Protection With Sprinkler Irrigation,*" University of Michigan Extension Bulletin 327, 1955.

Dickerson, W.H. and Dethier, B.E., "*Drought Frequency in the Northeastern United States,*" West Virginia University Agricultural Experiment Station Bulletin 595, June 1970.

Palmer, Wayne C., *Meteorological Drought,* U.S. Weather Bureau Research Paper No. 45, 1965.

Rogers, William J., and Swift, Harry L., *Frost and the Prevention of Frost Damage,* National Weather Service, National Oceanic and Atmospheric Administration, U.S. Department of Commerce, 1970.

## Weather and Food Production

**Articles:**

Aldrich, Samuel R., "Soil Thermometers," *Crops and Soils,* Jan. 1968.

Burnside, O.C., Fenster, C.R., Wicks, G.A., and Drew, J.U., "Effect of Soil and Climate on Herbicides Dissipation," *Weed Science,* Vol. 17, 1969, pp. 241-245.

Caprio, Joseph M., "Pattern of Plant Development in the Western United States," *Montana Agricultural Experiment Station Bulletin 607,* 1966.

Collis, R.T.H., "Weather and World Food," *Bulletin, American Meteorological Society,* Vol. 56, No. 10, Oct. 1975, pp. 1078-1083.

Davis, D. R., and Jensen, R. E., "Weather Forecast and Advisory Service for Agriculture in the Southeast," *Proceedings, Florida State Horticultural Society,* Vol. 87, 1974, pp. 149-155.

Hopp, Richard, J., and Vittum, M. T., "Nature's Own Weather Bureau," *Organic Gardening and Farming,* April, 1977.

Owen, P.C., and Watson, D. J., "Effect on Crop Growth of Rain After Prolonged Drought," *Nature,* Vol. 177, 1956, p. 847.

Penman, H. L., "Weather and Crops," *Quarterly Journal,* Royal Meteorological Society, Vol. 88, 1962, pp. 209-219.

Stirm, Walter, L., and Newman, James E., "Weather Equipment," *Crops and Soils,* Jan. 1968.

**Books:**

*Climate and Food: Climatic Fluctuation and U.S. Agricultural Production,* Report of the Committee on Climate and Weather

Fluctuations and Agricultural Production, National Academy of Sciences, Washington, D.C., 1976.

Taylor, James A., editor, *Weather and Agriculture,* Pergamon Press, New York, N.Y., 1967.

*Climate and Man: Yearbook of Agriculture,* U.S. Department of Agriculture, U.S. Government Printing Office, Washington, D.C., 1941.

## Climate and Climatic Change

**Articles:**

Alexander, T., "Ominous Changes in the World's Weather," *Fortune,* Vol. 89, No. 2, 1974, pp. 90-152.

"Canadian Climate Data Publications," *Weatherwise,* Vol. 24, No. 6, Dec. 1971, p. 287.

Landsberg, Helmut E., "Two Centuries of New England Climate," *Weatherwise,* Vol. 20, No. 2, April 1967, p. 52.

Ludlum, David M., "Early and Late Season Snowfalls: Evidence of Climatic Change?", *Weatherwise,* Vol. 21, No. 2, April 1968, p. 62.

Matthews, Samuel W., "What's Happening to Our Climate?" *National Geographic,* Vol. 150, No. 5, Nov. 1976, pp. 576-615.

Newman, James E., "Climatic Changes: Some Evidence and Implications," *Weatherwise,* Vol. 24, No. 2, April 1971, p. 56.

"Rise in Global Concentrations of Carbon Dioxide," *Weatherwise,* Vol. 30, No. 5, Oct. 1977, p. 207.

Stockton, Charles W., and Meko, David M., "A Long-Term History of Drought Occurrence in the Western United States as Inferred from Tree Rings," *Weatherwise,* Vol. 28, No. 6, Dec. 1975, p. 244.

Wahl, E. W., "A Comparison of the Climate of the Eastern United States During the 1830s with the Current Normals," *Monthly Weather Review,* Vol. 96, 1968, pp. 73-82.

**Books:**

Baldwin, John L., *Climates of the United States,* Environmental Data Service, National Oceanic and Atmospheric Administration, U. S. Department of Commerce, Washington, D.C., 1973.

Bryson, Reid A., and Murray, Thomas J., *Climates of Hunger,* University of Wisconsin Press, Madison, Wis., 1977.

Claiborne, Robert, *Climate, Man, and History,* W. W. Norton & Co., New York, N.Y., 1970.

Hays, James D., *Our Changing Climate,* Atheneum Publishers, N.Y., 1977.

Ponte, Lowell, *The Cooling,* Prentice-Hall, Inc., Englewood Cliffs, N.J., 1976.

Ruffner, James A., and Bair, Frank E., *Weather Almanac,* 2nd Ed., Gale Research Co., Detroit, Mich., 1978.

## Miscellaneous Books

Adams, Anthony, *Your Energy Efficient House,* Garden Way Publishing, Charlotte, Vt., 1977.

Crowther, Richard L., *Sun/Earth* (a guide to the use of solar and other natural energy systems for homes and buildings), American Meteorological Society, Boston, Mass., 1976.

Daniels, Farrington, *Direct Use of the Sun's Energy,* Ballantine Books, New York, N.Y., 1974.

# E
# SOURCES OF TECHNICAL INFORMATION

Some readers may be interested in obtaining more technical information on certain topics covered in this book. For those with the background and interest to read them, a list of pertinent scientific journals and books is presented below. Although the list is by no means complete, it does include most of the major journals in which articles in the general field of weather and agriculture are likely to be found.

Some of these journals and books may be located in local libraries, but it is likely that many will be found only in larger state libraries or at universities—especially those with programs in agriculture and/or meteorology. If you cannot locate a publication you desire, give the nearest state or university library a call and ask if it might be available there.

## Journals:

*Agricultural Engineering*
*Agricultural Meteorology*
*Agronomy Journal*
*Canadian Journal of Plant Science*
*Crops and Soils*
*Crop Science*
*Experimental Agriculture*
*Journal of Agricultural Science*
*Journal of Applied Meteorology*
*Journal of Horticultural Science*
*Monthly Weather Review*

*Nature*
*Plant and Soil*
*Plant Physiology*
*Quarterly Journal, Royal Meteorological Society*
*Soil Science*

## Books:

*Agricultural Meteorology,* Meteorological Monographs Vol. 6, No. 28, American Meteorological Society, Boston, Mass. July, 1965.

Chang, Jen-Hu, *Climate and Agriculture,* Aldine Publishing Co., Chicago, Ill., 1968.

Geiger, Rudolf, *The Climate Near the Ground,* Harvard University Press, Cambridge, Mass., 1965.

Leith, Helmut, *Phenology and Seasonality Modeling* (Ecological Studies 8), Springer-Verlag, New York, N.Y., 1975.

Levitt, J. *Responses of Plants to Environmental Stresses,* Academic Press, New York, N.Y., 1972.

Munn, Robert E., *Descriptive Micrometeorology,* Academic Press, New York, N.Y., 1966.

Slatyer, R.O., *Plant-Water Relationships,* Academic Press, London, England, 1967.

Wang, W., *Agricultural Meteorology,* Pacemaker Press, 1963.

Wang, Jen-Yu, and Barger, Gerald L., *Bibliography of Agricultural Meteorology,* University of Wisconsin Press, Madison, Wisc., 1962.

# F

# WEATHER DATA AND THE METRIC SYSTEM

Sometime in the not-too-distant future—perhaps even by the time you are reading this book—weather information in the United States will contain some strange sounding values. The weather is going metric! As part of an overall plan to transform our system of measurements in this country into the metric system, the National Weather Service has announced plans to begin using metric units in the near future. Already, temperatures are often given in both degrees Fahrenheit (°F.) and degrees Celsius (°C.). The exact date of a complete transfer to metric units by the NWS is uncertain and the schedule has already been pushed back at least a year by adverse public reaction. Unless further and persistant complaints are voiced by the general public, however, we will soon be measuring temperatures in degrees Celsius (°C.), precipitation in millimeters (mm), snowfall in centimeters (cm) and wind speed in meters per second (m/s).

Usually, the reason given for the necessity of this change is that virtually all other countries use metric units in reporting weather data and if the United States adopts that system it will facilitate the exchange of data among nations. Actually, for many years we have exchanged weather data with other countries in metric units, while maintaining our current system for domestic purposes. So the only real effect will be on you, the general public. One could question the need to change our system solely for domestic purposes, especially since there has been no noticeable demand for metric weather data, but the bureaucracy has decreed and it shall thus be! Any additional protests from the public are likely to merely delay, rather than cancel, the changeover. That being the case, let's look at the two systems and how they compare. (There is certainly no need to rush out to purchase weather instruments calibrated in metric units and no reason for you to change your current system of recording weather information unless you really want to. Published climatological data using °F. and inches are likely to be available for many years.)

## Temperature

Current units: degrees Fahrenheit (°F.)
Metric units: degrees Celsius (°C.)

*Note:* Celsius and Centigrade are essentially the same unit. Perhaps Celsius was chosen because the average person is generally more familiar with the term Centigrade—why make it too easy?

| | Freezing Point | Boiling Point |
|---|---|---|
| °F. | 32° | 212° |
| °C. | 0° | 100° |

To convert °F. to °C. : °C.=5/9 (°F.−32)
subtract 32 from the °F. value and multiply the result by 5/9, or 0.556.

To convert °C. to °F. : °F.=9/5C+32
multiply the °C. value by 9/5 (or 1.8) and then add 32 to the result.

For quick and easy conversion, Table F-1 is included. Simply look up the F. or C. value and then read the corresponding value in the other unit next to it.

## Table F-1
## Temperature Conversions Between °F. and °C.

| F | C | F | C | F | C | F | C | F | C | F | C |
|---|---|---|---|---|---|---|---|---|---|---|---|
| 100 | 38 | | | | | | | | | | |
| 99 | 37 | 79 | 26 | 59 | 15 | 39 | 4 | 19 | −7 | −1 | −18 |
| 98 | 37 | 78 | 26 | 58 | 14 | 38 | 3 | 18 | −8 | −2 | −19 |
| 97 | 36 | 77 | 25 | 57 | 14 | 37 | 3 | 17 | −8 | −3 | −19 |
| 96 | 36 | 76 | 24 | 56 | 13 | 36 | 2 | 16 | −9 | −4 | −20 |
| 95 | 35 | 75 | 24 | 55 | 13 | 35 | 2 | 15 | −9 | −5 | −21 |
| 94 | 34 | 74 | 23 | 54 | 12 | 34 | 1 | 14 | −10 | −6 | −21 |
| 93 | 34 | 73 | 23 | 53 | 12 | 33 | 1 | 13 | −11 | −7 | −22 |
| 92 | 33 | 72 | 22 | 52 | 11 | 32 | 0 | 12 | −11 | −8 | −22 |
| 91 | 33 | 71 | 22 | 51 | 11 | 31 | −1 | 11 | −12 | −9 | −23 |
| 90 | 32 | 70 | 21 | 50 | 10 | 30 | −1 | 10 | −12 | −10 | −23 |
| 89 | 32 | 69 | 21 | 49 | 9 | 29 | −2 | 9 | −13 | −11 | −24 |
| 88 | 31 | 68 | 20 | 48 | 9 | 28 | −2 | 8 | −13 | −12 | −24 |
| 87 | 31 | 67 | 19 | 47 | 8 | 27 | −3 | 7 | −14 | −13 | −25 |
| 86 | 30 | 66 | 19 | 46 | 8 | 26 | −3 | 6 | −14 | −14 | −26 |
| 85 | 29 | 65 | 18 | 45 | 7 | 25 | −4 | 5 | −15 | −15 | −26 |
| 84 | 29 | 64 | 18 | 44 | 7 | 24 | −4 | 4 | −16 | −16 | −27 |
| 83 | 28 | 63 | 17 | 43 | 6 | 23 | −5 | 3 | −16 | −17 | −27 |
| 82 | 28 | 62 | 17 | 42 | 6 | 22 | −6 | 2 | −17 | −18 | −28 |
| 81 | 27 | 61 | 16 | 41 | 5 | 21 | −6 | 1 | −17 | −19 | −28 |
| 80 | 27 | 60 | 16 | 40 | 4 | 20 | −7 | 0 | −18 | −20 | −29 |

## Precipitation

Current units: inches and tenths of inches (in)
Metric units: millimeters and centimeters (mm and cm)

There will be two units used in precipitation measurements—millimeters (mm) for total precipitation (rainfall and melted snow) and centimeters (cm) for snowfall. Conversion is fairly simple:

*1 inch=2.54 cm=25.4 mm*
*1 cm=0.39 inch*
*1 mm=0.039 inch*

*1 cm=10 mm*
*1 mm=0.1 cm*

For quick conversions a value of 2½ cm or 25 mm to the inch can be used. This will provide values suitable for most general purposes. Thus:

*2 inches=5 cm=50 mm*
*4 inches=10 cm=100 mm*

*8 inches=20 cm=200 mm*
*etc., etc., etc.,*

### Wind

Current units: miles per hour (MPH)
Metric units: meters per second (m/s)

To convert between units:

**1 m/s=2.24 mph** **1 mph=0.45 m/s**

Thus, as a rule-of-thumb, the wind speed in miles per hour is just over twice the value given in meters per second, while the m/s value is just under half of the MPH speed.

Some key conversions:

| m/s | MPH | MPH | m/s |
|---|---|---|---|
| 1 | 2.2 | 1 | 0.4 |
| 5 | 11.2 | 5 | 2.2 |
| 10 | 22.4 | 10 | 4.5 |
| 15 | 33.6 | 15 | 6.7 |
| 20 | 44.7 | 20 | 8.9 |
| 25 | 55.9 | 25 | 11.2 |
| 30 | 67.1 | 30 | 13.4 |
| 35 | 78.3 | 35 | 15.6 |
| 40 | 89.5 | 40 | 17.9 |

# G

# USE OF THE GRAVIMETRIC METHOD TO DETERMINE SOIL MOISTURE

Soil samples should be collected from several locations at the site you are interested in. Make sure you dig down several inches to obtain soil from the root zone of your crops as well as from the surface layer. Thoroughly mix the samples together and then select a small amount from this mixture for your test. You may want to make several tests and average the results together for a final value. First, weigh your clean and heat-proof container when it is empty and then again with the soil sample included. Heat the container and soil in an oven at about 100° to 110° C. (212° to 230°F.) until the soil is completely dry. Then weigh the container with the soil again.

Assume, for demonstration purposes, that your empty container weighs 20 oz., and the filled container weighs 30 oz. before heating and 27.5 oz. after heating. Then the soil moisture would be calculated as follows:

| | |
|---|---|
| Full container before heating | = 30 oz. |
| Empty container | = 20 oz. |
| Initial weight of soil sample | = 10 oz. |

| | |
|---|---|
| Full container after heating | = 27.5 oz. |
| Empty container | = 20.0 oz. |
| Weight of dry soil | = 7.5 oz. |

| | |
|---|---|
| Initial weight of soil sample | = 10.0 oz. |
| Weight of dry soil | = 7.5 oz. |
| Weight of water in soil | = 2.5 oz. |

Percentage (by weight) of soil moisture = 2.5÷7.5=33 per cent

# H

# ANALYSIS OF HOME WEATHER STATION DATA

If you have a home weather station and record daily values of rainfall or maximum/minimum temperatures you can easily calculate some simple climatological averages for your location, after you have a few years of data. For any meaningful averages you should have at least five years of records and 10 years would be better. The longer your record, the more representative it will be of your long-term climatic conditions.

The average of any series of values can be easily calculated by simply adding the values together to obtain their sum, or total, and dividing the sum by the number of individual values. Thus, if you have a 10-year record of growing season rainfall, to obtain your average rainfall during the growing season add the ten years of rainfall totals together and divide by ten.

Most home gardeners are interested in three basic weather parameters—rainfall, temperatures, and freeze dates. Sample analyses of these parameters follow.

## Rainfall

You can determine average rainfall for any time period you wish—your growing season, the full year, or each individual week or month. As an example, let's assume you have 12 years of precipitation records and are interested in calculating your average growing season rainfall. Further, assume you normally begin planting in April and that your garden is essentially finished by the end of September. Thus, you are interested in the total rainfall from April through September. Your average growing season rainfall is then calculated as follows:

| Year | | Total Rainfall April - Sept. |
|---|---|---|
| 1 | | 25.03 |
| 2 | | 29.98 |
| 3 | | 24.34 |
| 4 | | 18.87 |
| 5 | | 26.91 |
| 6 | | 17.55 |
| 7 | | 14.70 |
| 8 | | 19.36 |
| 9 | | 26.10 |
| 10 | | 23.35 |
| 11 | | 24.17 |
| 12 | | 21.85 |
| | Total | 272.21 |
| | Number of years | 12 |

Average growing season rainfall =272.21÷12=22.68 inches

Using the same technique you can calculate average rainfall for any period of interest. The individual values can be plotted on a graph. Plotting the data on a graph gives you a better picture of any trends that may be occurring. Each year you can update your average. After you have accumulated a long period of record you can calculate your average rainfall for, say, the most recent five growing seasons and compare it with your overall average. This comparison will give you an indication of whether your recent growing seasons have been wetter or drier than "normal."

## Temperature

Daily mean temperatures are determined by simply adding the daily maximum and minimum temperatures together and dividing by two. To calculate monthly averages simply add all the daily maximum temperatures and then all the daily minimums.

Divide these sums by the number of days in the month and you will have your average daily maximum and minimum temperatures for that month. For example, in October 1978, the sum of my daily maximum temperatures was 1987 and the sum of my daily minimums was 1292. Dividing each of these sums by the 31 days in the month gave me an average maximum temperature of 64.1° (1987÷31) and an average minimum of 41.7° (1292÷31). The overall average temperature for the month was (64.1+41.7)÷2 or 52.9°F.

Long-term monthly or annual average temperatures can be calculated in the same manner as shown for rainfall data. As an example, the average temperatures during August for the past six years at my garden site have been:

| Year | Average Temp. | |
|---|---|---|
| 1973 | 73.5 | |
| 1974 | 71.7 | |
| 1975 | 73.1 | |
| 1976 | 70.0 | |
| 1977 | 72.6 | |
| 1978 | 74.0 | |
| | Total=434.9 | 434.9÷6=72.5 |

Thus, my average August temperature for the six year period has been 72.5°F.

You may also want to keep track of extreme temperatures that occur each month or for some other period. Again using my records for August as an example:

| Year | Highest Temp. | Lowest Temp. |
|---|---|---|
| 1973 | 95 | 49 |
| 1974 | 84 | 57 |
| 1975 | 93 | 53 |
| 1976 | 87 | 45 |
| 1977 | 90 | 48 |
| 1978 | 89 | 53 |
| Extremes for period: | 95 | 45 |

Such data are useful to determine the range of extreme conditions you are likely to have to contend with in any given year.

## Freeze Data

Recording the last 32° or lower temperature in spring and the first 32° or lower temperature in the fall is quite simple with a max/min thermometer. In order to calculate average freeze dates from these data you will first need to convert the various calendar dates to a day number. The simplest scheme is to start with January 1st as Day 1 and continue throughout the year. Thus, February 1st becomes Day 32, March 1st is Day 60 (61 in leap years), etc. Table H-1 lists the Day # for each day of the year. Note that in leap years you should add one to each Day # beginning with March 1st, since February 29 becomes Day 60.

After converting your freeze dates to day numbers you can then calculate average values for both spring and fall freeze dates in the same way you calculated average temperatures. Example:

| Year | Last 32° in Spring | Day # | First 32° in Fall | Day # | Freeze-free Period (days) |
|---|---|---|---|---|---|
| 1 | April 15 | 105 | Oct. 10 | 283 | 178 |
| 2 | April 20 | 110 | Nov. 6 | 310 | 200 |
| 3 | April 22 | 112 | Oct. 3 | 276 | 164 |
| 4* | April 13 | 104 | Oct. 31 | 305 | 201 |
| 5 | April 29 | 119 | Oct. 19 | 292 | 173 |
| 6 | May 1 | 121 | Oct. 18 | 291 | 170 |
| 7 | April 21 | 111 | Oct. 18 | 291 | 180 |
| 8* | April 23 | 114 | Oct. 21 | 295 | 181 |
| | Sum= | 896 | | 2343 | 1447 |

*Leap years

| | | | |
|---|---|---|---|
| **# Years record=** | **8** | **8** | **8** |
| **Average** | **112** | **293** | **181** |
| **Calendar day** | **April 22** | **Oct. 20** | |

After the average Day # values have been calculated for the freeze dates, use Table H-1 to convert back to calendar days. In this example, the average date of the last 32° in spring is April 22 and the average date of the first 32° in fall is October 20.

The above example also lists the length of the freeze-free period each year (Day # of first freeze fall freeze minus Day # of last spring freeze) and the average for the period of record—181 days in this case. All of these values (freeze dates and length of the freeze-free period) can be plotted on graph paper to give a visual presentation of current trends and variations.

## Table H-1: Conversions from Calendar Dates to Day Number
(January 1 Is Day Number 1)

| Feb. | Mar. | Apr. | May | June | July | Aug. | Sept. | Oct. | Nov. | Dec. |
|---|---|---|---|---|---|---|---|---|---|---|
| 1-32 | 1-60 | 1-91 | 1-121 | 1-152 | 1-182 | 1-213 | 1-244 | 1-274 | 1-305 | 1-335 |
| 2-33 | 2-61 | 2-92 | 2-122 | 2-153 | 2-183 | 2-214 | 2-245 | 2-275 | 2-306 | 2-336 |
| 3-34 | 3-62 | 3-93 | 3-123 | 3-154 | 3-184 | 3-215 | 3-246 | 3-276 | 3-307 | 3-337 |
| 4-35 | 4-63 | 4-94 | 4-124 | 4-155 | 4-185 | 4-216 | 4-247 | 4-277 | 4-308 | 4-338 |
| 5-36 | 5-64 | 5-95 | 5-125 | 5-156 | 5-186 | 5-217 | 5-248 | 5-278 | 5-309 | 5-339 |
| 6-37 | 6-65 | 6-96 | 6-126 | 6-157 | 6-187 | 6-218 | 6-249 | 6-279 | 6-310 | 6-340 |
| 7-38 | 7-66 | 7-97 | 7-127 | 7-158 | 7-188 | 7-219 | 7-250 | 7-80 | 7-311 | 7-341 |
| 8-39 | 8-67 | 8-98 | 8-128 | 8-159 | 8-189 | 8-220 | 8-251 | 8-281 | 8-312 | 8-342 |
| 9-40 | 9-68 | 9-99 | 9-129 | 9-160 | 9-190 | 9-221 | 9-252 | 9-282 | 9-313 | 9-343 |
| 10-41 | 10-69 | 10-100 | 10-130 | 10-161 | 10-191 | 10-222 | 10-253 | 10-283 | 10-314 | 10-344 |
| 11-42 | 11-70 | 11-101 | 11-131 | 11-162 | 11-192 | 11-223 | 11-254 | 11-284 | 11-315 | 11-345 |
| 12-43 | 12-71 | 12-102 | 12-132 | 12-163 | 12-193 | 12-224 | 12-255 | 12-285 | 12-316 | 12-346 |
| 13-44 | 13-72 | 13-103 | 13-133 | 13-164 | 13-194 | 13-225 | 13-256 | 13-286 | 13-317 | 13-347 |
| 14-45 | 14-73 | 14-104 | 14-134 | 14-165 | 14-195 | 14-226 | 14-257 | 14-287 | 14-318 | 14-348 |
| 15-46 | 15-74 | 15-105 | 15-135 | 15-166 | 15-196 | 15-227 | 15-258 | 15-288 | 15-319 | 15-349 |
| 16-47 | 16-75 | 16-106 | 16-136 | 16-167 | 16-197 | 16-228 | 16-259 | 16-289 | 16-320 | 16-350 |
| 17-48 | 17-76 | 17-107 | 17-137 | 17-168 | 17-198 | 17-229 | 17-260 | 17-290 | 17-321 | 17-351 |
| 18-49 | 18-77 | 18-108 | 18-138 | 18-169 | 18-199 | 18-230 | 18-261 | 18-291 | 18-322 | 18-352 |
| 19-50 | 19-78 | 19-109 | 19-139 | 19-170 | 19-200 | 19-231 | 19-262 | 19-292 | 19-323 | 19-353 |
| 20-51 | 20-79 | 20-110 | 20-140 | 20-171 | 20-201 | 20-232 | 20-263 | 20-293 | 20-324 | 20-354 |
| 21-52 | 21-80 | 21-111 | 21-141 | 21-172 | 21-202 | 21-233 | 21-264 | 21-294 | 21-325 | 21-355 |
| 22-53 | 22-81 | 22-112 | 22-142 | 22-173 | 22-203 | 22-234 | 22-265 | 22-295 | 22-326 | 22-356 |
| 23-54 | 23-82 | 23-113 | 23-143 | 23-174 | 23-204 | 23-235 | 23-266 | 23-296 | 23-327 | 23-357 |
| 24-44 | 24-83 | 24-114 | 24-144 | 24-175 | 24-205 | 24-236 | 24-267 | 24-297 | 24-328 | 24-358 |
| 25-56 | 25-84 | 25-115 | 25-145 | 25-176 | 25-206 | 25-237 | 25-268 | 25-298 | 25-329 | 25-359 |
| 26-57 | 26-85 | 26-116 | 26-146 | 26-177 | 26-207 | 26-238 | 26-269 | 26-299 | 26-330 | 26-360 |
| 27-58 | 27-86 | 27-117 | 27-147 | 27-178 | 27-208 | 27-239 | 27-270 | 27-300 | 27-331 | 27-361 |
| 28-59 | 28-87 | 28-118 | 28-148 | 28-179 | 28-209 | 28-240 | 28-271 | 28-301 | 28-332 | 28-362 |
| | 29-88 | 29-119 | 29-149 | 29-180 | 29-210 | 29-241 | 29-272 | 29-302 | 29-333 | 29-363 |
| | 30-89 | 30-120 | 30-150 | 30-181 | 30-211 | 30-242 | 30-273 | 30-303 | 30-334 | 30-364 |
| | 31-90 | | 31-151 | | 31-212 | 31-243 | | 31-304 | | 31-365 |

# Index


A

Accuracy of weather forecast, 116
Advection freeze, 68, 74, 78
Agricultural drought:
  definition, 82
Air pollution, 77, 153
Air pressure;
  measurement, 11 and *illus.*; *see also* Barometer
Air temperature:
  diurnal change, 37
  effect on plant disease, 60, 61
  effect on plant growth, 34–38, 96, *illus.* 98, 126; *see also* Plants: temperature
  measurement, 1, 2, 4–5, 6, 16: *see also* Maximum/minimum thermometer; Thermograph (air)
  microclimate, 130–33
  in phenology, 23
  records, 16, *illus.* 17, *illus.* 18
  and transpiration, 51
  vertical differences, 131–33
Apple:
  blossom: dates, 100, *illus.* 101; as indicator plant, 25
  critical temperatures, 71
  scab (disease), 64
  stomates, 50
Available water, 53

B

Bacterial wilt of sweet corn, 62–63
Barometer, 11 and *illus.*
Base temperature, 103
Bean(s):
  critical temperatures, 71
  moisture-sensitive period, 52
  soil temperature, 41
  stomates, 50
  temperature requirements, 41 and *illus.*
Bioclimatic law, 22–23
"Black frost," 68
Blight (plant disease), 58–59, 63
Blossom end rot, 58–59
Brassicas, 25

C

Calculated data for weather records, 18
Calendar days to maturity (plants), 95–96, 103, 126
Carbon dioxide:
  effect on air temperature, 145, 152, 153
  effect on plant growth, 54–55
  in photosynthesis, 44, 54
Cardinal temperature of plants, 37–38, 70, 71 (*Tables 8, 9*), 96, 124
Climate:
  change, 145–157; causes, 150–53
  definition, 120
  history, 147–49
Climatological shelter, *see* Weather shelter
Climatology, 119–27
Cloche, frost protection), *illus.* 74
CMI, *see* Crop Moisture Index
Cold frame, 70 and *illus.*
Cold hardiness, 70
Convection currents, 130
Corn:
  bacterial wilt, 62–63
  critical temperature, 71
  efficiency, 45
  growing degree days to maturity, *illus.* 98–99, 126
  leaf blight, 65
  moisture requirements, 36, 52, *illus.* 86
  planting time, 25
  solar energy requirements, 46
  stomates, 50
  temperature requirements: air, 36 and *illus.*; soil, 39
Cotton:
  diseases, 65
  temperature requirements, 39
Cranberry:
  frost protection, 77
Critical temperature of plants, 37–38, 70, 71 (*Tables 8, 9*), 96, 124
Crop development:
  definition, 33
Crop distribution:
  effects of weather, 35–36
Crop Growth:
  effects of weather, 33–55, 96, *illus.* 98
Crop Moisture Index, 53–54, 86

Crop record (phenological), 104–106
Crop site selection, 26, 132–33
Curly top (virus plant disease), 64

D

Daylength, *see* Photoperiodism
Dew:
  measurement, 10
  and plant disease, 61
Dew-point temperature, 9, 80
Disease (plant), *see* Plants: disease
Dormancy requirements of plants, 35
Downy mildew, 58–59, 62
Drip irrigation, 85 and *illus.*
Drought, 52–53, 80–86, 154
  agricultural: definition, 82
  forecasting, 86
  meteorological: definition, 82
  protection, 49, 82–85, 88
  severity, 86

E

Efficiency of plants, 45–46
Energy efficiency of house, 141–43
Ephemeral plants, 52
Equipment, *see* Instruments
Evaporation, 44
  measurement, 9, 16, 85
Evaporimeter, 9
Evapotranspiration, 51 and *illus.*, 53, 81, 82, 140

F

FC, *see* Field capacity
Field capacity, 53, 85
Forecasting:
  drought, 86
  frost, 9, 78–80
  plant disease, 62–65
Forecast of weather:
  accuracy, 116
  interpretation, 109–17
  long-range, 119–20
Forsythia:
  as indicator plant, 25, 29
Freeze, *see* Frost
Frost, 67–80
  damage to plants, 38, 69
  dates: fall, 122 and *map*, 156; local data, *map* 123; spring, 121 and *map*, 156
  forecasting, 9, 78–80
  hardiness, 70
  pocket, 139
  probability, 120–23, 126
  protection, 72–77
Fruits:
  critical temperatures, 71 (*Table 9*)
  storage, 55
Fungicide:
  dust, 61
Fungus plant diseases, 58, 60
Fusarium wilt, 42, 58–59

G

Garden site selection, 138–40 and *illus.*
GDD, *see* Growing degree days
Grain crops:
  temperature requirements, 36
Green bean(s), *see* Bean(s)
"Greenhouse effect" in atmosphere, 152
Growing degree days, 18, 97–106, 126
  formula, 97–98
  modified formula, 99–100
  record-keeping, 104
Growing season:
  definition, 123–24

H

Hail, 90–91, 126–27
HDD, *see* Heating Degree Days
Heater (frost protection), 75
Heating degree days, 18, 106–107 and *map*
  formula, 106
Helicopter (frost protection), 75–76
Hoar frost, *see* Frost
Home:
  energy efficiency, 141–43
  site selection, 141–42
  weather station, 1–18, 127
Honeysuckle:
  as indicator plant, 23, 27
Hopkins, A. D., 22
Hopp, Richard, 25
Humidity, *map* 63, 127
  measurement, 9–10
  and plant disease, 61
and transpiration, 51
Hydrologic cycle, *illus.* 50
Hydrothermograph, 6, 10, 80

I

Ice age, 146, 147, 155, 156; *see also* Little Ice Age
Indicator plants, 23–24, 25–29 *passim*
  criteria, 27
Insecticide:
  dust, 61
Instruments (weather), 3–11
  location, 14–15
  non-recording, 3, 4–5, 7–11 *passim*
  recording, 3, 6–7
  shelter, *see* Weather shelter
Interpretation of weather forecast, 109–17
Inversion, 75, 130–31
Irrigation, 82, 84–85, 125, 133
  drip method, 85 and *illus.*

L

Lake breeze, *illus.* 135
Landscaping, 142–43
Langley (calories per square centimeter), 47–48
Leaf temperature, *see* Plants: temperature
Lethal temperature of plants, 37
Lettuce:
  temperature requirements, 34–35 and *illus.*, 37
Light, *see* Sunlight
Lilac:
  bloom dates, 100, *illus.* 101
  as indicator plant, *map* 22, 23, *map* 24, 25, *illus.* 26, 27, *illus.* 28, *illus.* 29
  Solar Thermal Unit requirement, 100
Little Ice Age, 149, 155
Location, *see* Site selection
Long-range weather forecast, 119–20

M

Macroclimate:
  definition, 129
Maturity of plants:
  measurement, 95–96, 103, 126
Maximum/minimum thermometer, 2 and *illus.*, 3, 4 and *illus.*, 5, 6, 97, 104, 133, 156
Measurement of:
  air pressure, 11 and *illus.*
  air temperature, 1, 2, 4–5, 6, 16, 17, 18
  days to maturity (plants), 95–96, 103, 126
  dew, 10
  evaporation, 9, 16
  rainfall, 2, 5, 6–7, 16, 17, 18
  relative humidity, 9–10
  soil moisture, 8 and *illus.*, 9
  soil temperature, 7–8, 16, 17
  solar radiation10–11
  wind, 10 and *illus.*
Melon(s):
  critical temperatures, 38 (*Table 3*)
  temperature requirements, 37
Mesoclimate:
  definition, 129
Meteorological drought:
  definition, 82
Meterorologist, 114–15
Meteorology:
  definition, 120

Microclimate, 129–43
air temperature, 130–33
definition, 129
precipitation, 133–34
wind, 135
Microwave radiation (frost protection), 77
Mildew (plant diseases), 58–59
Minimum/maximum thermometer, *see* Maximum/minimum thermometer
Moisture, *see* Water
Moisture meter (soil), 8 and *illus.*, 9
Mulch:
as drought protection, 83
and frost, 69, 73

N

Narrow row planting, 83
National Climatic Center, Asheville, N.C., 127
National Oceanic and Atmospheric Administration (NOAA):
radio transmitters, *map* 117
National Weather Service, 23, 68, 114–15, 127
Nematode, 42, 58–59, 60, 65
NOAA Weather Radio. 116, *map* 117
Non-recording instruments (weather), 3, 4–5, 7–11 *passim*
Northeast Lilac Phenology Network, *map* 24
NWS, *see* National Weather Service

O

Onion, 35
moisture-sensitive period, 52
(*Table 7*)
smut (disease), 65
Optimum temperature of plants, 37
Orchard:
site selection, 138, 140–41
Organic matter added to soil, 82, 139

P

Palmer, Wayne, 54, 86
Palmer Index (of drought conditions), 86
Pea(s):
Growing degree day calculations, 101–103, 106
moisture-sensitive period, 52 (*Table 7*)
Pecan:
scab (disease), 64
Permanent wilting point, 53
Pest control:
biological, 61
Pesticide:
chemical, 61
Phenology, 21–29
network of observation sites, 23
record-keeping, 27, 29, 104–106, 156
Photoperiodism, 45, 46, 47–48
Photosynthesis, 44–45, 46, 49
Plants:
daylength requirements, 45, 46, 47–48
days to maturity, 95–96, 103, 126; *see also* Growing degree days
disease, 42, 57–65; *see also* individual diseases
drought-resistant, 52
efficiency, 45–46
ephemeral, 52
frost damage, 38
growth: effects of temperature on, 34–42, 96, *illus.* 98, 126; effects of weather on, 33–55
hardiness, 70
lethal temperature, 37
moisture stress, 52–53
temperature, 38–39
transpiration, 44, 50–51; *see also* Evapotranspiration
turgor, 49
wilting, 53, 81; *see also* Permanent wilting point
wind damage, 87–88
*see also* Vegetables
Pollution, 77, 153
Potato:
blight, 58–59, 63
critical temperature, 71
moisture sensitive period, 52 (*Table 7*)
soil temperature, 40
water usage, 49 (*Table 6*)
Powdery mildew, 58–59, 61
Precipitation:
effect of terrain on, 133–34 and *illus.*
measurement, 2, 5, 6–7, 16, 17, 133, *illus.* 157
microclimate, 133–34
probability, 124–25
in weather forecast, 112
Prediction, *see* Forecasting
Psychrometer, 9 and *illus.*, 10, 80
Purple lilac, *see* Lilac

R

Radiant energy, *see* Solar radiation
Radiation:
solar, *see* Solar radiation
Radiation frost, *see* Frost
Rainfall:
and drought, 82
effect of terrain on, 133–34 and *illus.*
intensity, 6–7, 16, 85
measurement, 2, 5, 6–7, 16, 17, 18, 133, *illus.* 157
statistics, 124–25 and *map*
Raingage, 2, 3, 5 and *illus.*, 16, 133, 156
location, 14
plastic, 18
Recording instruments, 3, 6–7
Record-keeping:
growing degree days, 104, 106
phenology, 27, 29, 104–106, 156
weather, 1, 16–18, 27, 127, 156–57
Relative humidity, see Humidity
Root knot, 42, 65
Root rot, 58–59

Sea breeze, *illus.* 135
Selection of site(s), *see* Site selection
Shelter:
weather, 12–14 and *illus.*, 131; location, 14–15 and *illus.*
Shelterbelt, *see* Windbreak
Site selection:
for crops, 26, 132–33
garden, 138–40 and *illus.*
home, 141–42
orchard, 138, 140–41
Sling psychrometer, 9 and *illus.*, 10
Smoke screen (frost protection), 77
Snap bean(s), *see* Bean(s)
Snow depth, 17
Snowfall, 17, *illus.* 56
Soil:
moisture, 53–54; and drought, 82, measurement, 8 and *illus.*, 9, 85; and nematodes, 60; and transpiration, 51; *see also* Available water; Crop Moisture Index; Field capacity; Irrigation
organic matter added, 82, 139
temperature, 39–42; measurement, 7–8, 16, 17, 42; and nematodes, 60; *see also* Thermograph (soil)
Solar constant, 48
Solar radiation, 2, 42–45, 46, 100, 153
measurement, 10–11, 47–48
spectrum, 43 and *illus.*, 45
and transpiration, 51
wavelength, 42, 43 and *illus.*
*see also* Solar Thermal Unit; Sunlight
Solar Thermal Unit, 100–101
Sprinkler (frost protection), 76–77
Stewart's disease, 62–63

Stomate, 49, 50–51, 55
Strawberry:
daylength requirements, 47
irrigation, 84
String bean(s), *see* Bean(s)
STU, *see* Solar Thermal Unit
Succulants, 52
Sunlight, 42
in home design and orientation, 141, 142 and *illus.*
in phenology, 23
in photosynthesis, 44

T

Temperature:
air, *see* Air temperature
base, 103
critical, *see* Critical temperature of plants
of plants, *see* Plants: Temperature
profile, 131 and *illus.*
soil, see Soil: Temperature
Thermograph (air), 6 and *illus.*, 16
Thermograph (soil), 8
Thermometer:
maximum/minimum, 2 and *illus.*, 3, 4 and *illus.*, 5, 6, 97, 104, 133, 156
soil, 7–8 and *illus.*, 16, 17
Tobacco:
blue mold (disease), 64
Tomato:
critical temperature
diseases, 42, 58–59, 65
moisture-sensitive period, 52 (*Table 7*)
stomates, 50
temperature requirements: air, 37, 69; soil, 39, 40–41
transplant timing, 26
water usage, 49 (*Table 6*)
Transpiration, 44, 50–51
*see also* Evapotranspiration
Turgor, 49

U

U.S. Weather Bureau, *see* National Weather Service

V

Vegetables:
critical temperatures, 37–38, 71 (*Table 8*)
daylength requirements, 46 (*Table 5*), 47
moisture-sensitive periods, 52 (*Table 7*)
storage, 55
temperature requirements: air, 34–38; soil, 39–42
water requirements, 49–54
*see also* Plants; individual vegetables, e.g. Tomatoes
Vernalization, 35
Vertical temperature differences, 131–33
Verticillium wilt, 42, 58–59
Virus plant diseases, 58–59, 60–61

W

Warning:
definition, 113
Watch:
definition, 112
Water:
effect on plant disease, 60
effect on plant growth, 49–54; *see also* Precipitation; Rainfall; Humidity
Watering, *see* Irrigation
Weather:
definition, 120
Weather Bureau, *see* National Weather Service
Weather/crop relationship, 33–55
Weather forecast:
accuracy, 116
interpretation, 109–17
long-range, 119–20
Weather instruments:
location, 14–15
Weatherman, 114–15
Weather records, 1, 16–18, 27, 127, 156–57
calculated data, 18
*see also* Instruments
Weather shelter, 12–14 and *illus.*, 131
location, 14–15 and *illus.*
Weather station (home), 1–18, *illus.* 13
Weather vane, 10
Weather warning:
definition, 113
Weather watch:
definition, 112
*Weekly Weather and Crop Bulletin*, 54, 86, 127
Wheat stem rust, 64–65
White frost, *see* Frost
Wilting, 53, 81
*see also* Permanent wilting point
Wind, 87–89
direction, 111, 127
effect on microclimate, 131, 135
effect on plant disease, 61
machines (frost protection), 75–76 and *illus.*
measurement, 10, 88
and transpiration, 51
Windbreak, 88, *illus.* 89, 140, 143, 144 and *illus.*
Winter-kill, 69

# Other Garden Way Books You Will Enjoy

*The Complete Guide to Growing Berries & Grapes*, by Louise Riotte. 142 pages, quality paperback, $4.95. What to plant where, when, and exactly how.

*Profitable Herb Growing at Home*, by Betty E.M. Jacobs. 240 pages, quality paperback, $5.95. The perfect book for those who wish to expand a home herb garden into a moneymaking country sideline.

*Secrets of Companion Planting for Successful Gardening*, by Louise Riotte. 226 pages, quality paperback, $5.95, hardcover, $8.95. For bigger, more luscious crops.

*Down-to-Earth Vegetable Gardening Know-How*, featuring Dick Raymond, 160 pages, deluxe illustrated paperback, $5.95. Special durable cover edition, $7.95. A treasury of complete vegetable gardening information.

*Keeping the Harvest: Home Storage of Vegetables & Fruits*, by Nancy Thurber and Gretchen Mead. 208 pages, deluxe illustrated paperback, $5.95; spiral edition, $6.95; hardcover, $12.95. The very best of the food storage books.

*Vegetable Gardening Handbook*, by Roger Griffith. 120 pages, spiral bound, $3.95. Take it into your garden, for information and as your own record book.

*Dwarf Fruit Trees for the Home Gardener*, by Lawrence Southwick. 118 pages, quality paperback, $3.95; hardcover, $5.95. All you need to know to start a home orchard on a small plot.

*Let It Rot!*, by Stu Campbell. 152 pages, quality paperback, $3.95. Homemade fertilizers for a healthier garden.

*Cash from Your Garden*, by David Lynch. 208 pages, $5\frac{1}{2} \times 8\frac{1}{2}$, illustrated, quality paperback, $3.95. The book that tells how to turn your garden produce into extra income.

*The Home Gardener's Cookbook,* by Marjorie Blanchard. 192 pages, quality paperback, $4.95. Mouth-watering recipes using your garden produce.

*Treasured Recipes from Early New England Kitchens,* by Marjorie Blanchard. 144 pages, quality paperback, $4.95. Yesterday's favorite recipes, adapted to today's kitchens.

*Woodstove Cookery: At Home on the Range,* by Jane Cooper. 204 pages, illustrated by Sherry Streeter, $5.95. A warm and friendly introduction to buying, installing, and cooking on a kitchen wood-burning range, with hundreds of time-tested recipes.

*Homemade: 101 Easy-to-Make Things for Your House, Garden & Farm,* by Ken Braren and Roger Griffith. 176 pages, deluxe illustrated paperback, $6.95. A wonderful collection of simple projects for the home carpenter.

*Harnessing Water Power for Home Energy,* by Dermot McGuigan. 112 pp., quality paperback, $4.95, An authoritative, detailed look at the uses of water power for small-scale operations.

*Harnessing Wind Power for Home Energy,* by Dermot McGuigan. 144 pp., quality paperback, $4.95, library hardcover, $9.95. A thorough review of wind power systems for the homeowner, with illustrations and manufacturers directory.

*Low-Cost Pole Building Construction,* by Douglas Merrilees and Evelyn Loveday. 118 pp., deluxe paperback, $5.95. This will save you money, labor, time and materials.

These books are available at your bookstore, or may be ordered directly from Garden Way Publishing, Department 171X, Charlotte, VT 05445. If order is less than $10, please add 60¢ postage and handling.